Arduino Programming

A Beginner's Journey into Coding, Robotics, and Smart Tech Projects

by Maxwell Harper

Table of Contents

Introduction 3

Chapter 1: Introduction to Arduino Programming 8

Chapter 2: Understanding the Basics: What is Arduino and Why It's a Game-Changer 29

Chapter 3: Your First Project – Blinking an LED and Understanding Code 47

Chapter 4: Exploring Sensors and Actuators 62

Chapter 5: Coding Fundamentals: How Arduino Makes Coding Easy for Everyone 78

Chapter 6: Introduction to Robotics: The Fascinating World of Arduino Robots 98

Chapter 7: Building Smart Projects: From Concept to Creation 118

Chapter 8: Advanced Arduino Projects: Pushing the Limits of Your Skills 142

Chapter 9: Troubleshooting Like a Pro: Diagnosing and Solving Common Issues 163

Chapter 10: Moving Forward: How to Keep Learning and Innovating with Arduino 183

Chapter 11: Bonus Chapter – Arduino and the Future: How This Technology is Shaping Tomorrow 206

Conclusion 229

References 233

About the Author 237

Introduction

Welcome to Your Journey into Arduino Programming

Imagine this: You're holding a small device in your hands—an Arduino. To some, it might look like just a tiny circuit board, but to you, it's the key to unlocking a world of endless possibilities. Whether you're interested in robotics, smart tech, or simply want to understand how the digital world works behind the scenes, Arduino programming is your gateway to transforming ideas into reality.

Why Arduino?

Arduino isn't just for engineers or tech enthusiasts. It's for anyone with a curiosity about how technology works, anyone who has ever wondered how everyday gadgets like smart thermostats, automated lights, or even robot vacuums come to life. Arduino simplifies the process of learning to code and build projects, offering a user-friendly approach to the world of electronics and programming.

What sets Arduino apart is its community-driven, open-source nature. Unlike many other platforms, Arduino invites creators from all walks of life to experiment, share, and innovate. This makes it perfect for beginners who want to dive in without feeling overwhelmed by complex systems.

Who is This Book For?

This book is designed specifically for beginners—those with little to no programming experience. If you've never written a line of code or soldered a circuit, don't worry. We'll start slow and steady, building your skills with each chapter. By the end, you'll not only understand the basics of coding, but you'll be able to create interactive projects that combine hardware and software seamlessly.

You don't need any prior experience in electronics or robotics. You just need a willingness to learn, an eagerness to experiment, and the courage to fail—because failure is just another step toward mastery. We'll guide you through every step, ensuring you never feel lost or left behind.

What Will You Learn?

Throughout this book, we'll introduce you to the exciting world of Arduino, starting with the very basics and gradually moving toward more advanced topics. Here's a sneak peek at what you'll be doing:

- **Mastering the Basics of Coding**: You'll learn the fundamentals of programming, such as variables, loops, and conditions. We'll break everything down into digestible chunks, making it easier than ever to understand how code works.
- **Building Your First Projects**: We won't just teach you theory—we'll guide you through hands-on projects that you can start building right away. You'll begin by making an LED blink, then move on to interactive gadgets, sensors, and even basic robots.
- **Understanding How Sensors Work**: We'll introduce you to sensors—devices that allow your Arduino to interact with the real world. You'll learn how to use sensors for everything from measuring light levels to detecting motion.
- **Coding and Robotics**: We'll take a deeper dive into robotics, showing you how to build simple robots that

can follow lines, avoid obstacles, and more. You'll even learn how to control your projects wirelessly using Bluetooth or Wi-Fi!
- **Creating Smart Projects**: Whether you want to automate your home, build a weather station, or make your own smart devices, we'll show you how to integrate Arduino with other tech to make your projects truly "smart."
- **Troubleshooting Like a Pro**: You'll gain the skills to fix common errors, whether they are coding bugs or hardware issues. Arduino is all about trial and error, and you'll learn how to troubleshoot problems with confidence.
- **Future-Proofing Your Skills**: At the end of this journey, you'll not only have the skills to create your own projects but also the foundational knowledge to dive deeper into more complex fields like robotics, IoT (Internet of Things), and even machine learning.

A Practical, Step-by-Step Approach

This book is structured with one goal in mind: to make learning Arduino as enjoyable and practical as possible. We'll take a hands-on approach, starting with small, manageable projects that gradually increase in complexity. Each chapter builds upon the previous one, reinforcing what you've learned while introducing new concepts and techniques.

Along the way, you'll find detailed explanations, code examples, circuit diagrams, and tips that ensure you truly understand how things work. And, we don't just throw you into the deep end; you'll have plenty of time to experiment, make mistakes, and learn from them.

Why This Book is Different

There are plenty of Arduino books out there, but most of them overwhelm you with technical jargon or assume prior knowledge you don't yet have. What makes this book unique is our commitment to **simplicity** and **clarity**. We believe learning should be fun, accessible, and, most importantly, rewarding. Each chapter has been carefully crafted to ensure that no step is skipped, and no concept is too difficult to understand.

In addition, we take a holistic approach by focusing not only on the technical aspects but also on the **creative** side of Arduino. After all, Arduino isn't just about writing lines of code or connecting wires—it's about bringing your ideas to life. We'll encourage you to think outside the box, take on challenges, and approach each project as an opportunity for creative expression.

Are You Ready?

So, are you ready to embark on your Arduino adventure? This book will take you from beginner to creator in no time, helping you unlock a world of opportunities in coding, robotics, and smart technology.

By the end of this book, you'll be able to design and build your own Arduino projects, solve problems with confidence, and perhaps even inspire others with your work. You're about to join a global community of makers, tinkerers, and innovators who are changing the world one Arduino project at a time.

Let's get started. The future is in your hands, and it's waiting for you to bring it to life.

This introduction sets the tone for an engaging, inspiring learning experience and emphasizes practical, hands-on learning—elements that will resonate with beginners and differentiate your book from others on the market.

Chapter 1: Introduction to Arduino Programming

Your Gateway to the World of Coding and Innovation

Welcome to the world of Arduino programming! You've just taken the first step toward unlocking an exciting, creative, and incredibly rewarding journey into the realm of coding, electronics, and technology. Whether you've always been curious about how gadgets work or are seeking to develop your own projects, Arduino is the perfect tool for making your ideas come to life. This chapter will introduce you to the Arduino platform, explain what it is, and guide you through the initial steps to start your very first project. By the end of this chapter, you'll understand what Arduino is all about, and you'll be ready to dive into your first piece of code.

What is Arduino?

At its core, Arduino is an open-source electronics platform based on simple software and hardware. It's designed to make it easy for anyone—from students and hobbyists to professionals—to create interactive projects that involve both coding and physical devices (such as sensors, motors, lights, and more). Arduino has gained immense popularity because of its simplicity, flexibility, and accessibility, making it the go-to tool for anyone wanting to learn about coding and robotics.

Here's what makes Arduino so special:

- **Open Source**: Arduino is free to use, and the designs for both the hardware and software are open for

anyone to modify and improve. This has helped create a vibrant community of makers who share their projects, code, and resources.
- **Beginner-Friendly**: Whether you're a complete newbie or a seasoned programmer, Arduino is designed to be approachable. The programming language used by Arduino is a simplified version of C/C++, which makes it easy to learn.
- **Widely Supported**: There are endless tutorials, books, online forums, and user groups where you can find help, share your projects, and get inspired by others' creations.
- **Hardware and Software Integration**: Arduino allows you to write code that interacts directly with hardware, meaning that you can build physical projects—such as a robotic arm or a smart light system—by writing code.

Why Learn Arduino?

You may be wondering: **Why should I learn Arduino programming?** The answer is simple: Arduino is a powerful tool that opens up a world of possibilities for learning and creating. Here are just a few reasons why learning Arduino is so valuable:

1. **Hands-On Learning**: Unlike theoretical programming or coding, Arduino combines both hardware and software. You'll get to work with real-world projects, which makes the learning process interactive and highly engaging.
2. **Versatility**: Arduino isn't limited to one field. Whether you're into robotics, home automation, environmental monitoring, or gaming, Arduino can be used for a

wide variety of applications. If you have a vision, Arduino can help you bring it to life.
3. **Problem-Solving Skills**: The more you work with Arduino, the better your problem-solving skills will become. You'll learn to troubleshoot issues, debug your code, and solve technical challenges as you create your projects.
4. **Creativity and Innovation**: Arduino encourages creativity. With Arduino, you're not just learning to code—you're designing solutions to real-world problems and experimenting with new ideas. From building robots to developing smart devices, your potential is limited only by your imagination.
5. **Job and Career Opportunities**: In today's tech-driven world, understanding coding and electronics is an invaluable skill. Arduino can help you get started in the field of electronics, embedded systems, or robotics, and provide a strong foundation for more advanced topics like IoT (Internet of Things), AI (Artificial Intelligence), and machine learning.

The Arduino Hardware

Before we dive into coding, it's essential to understand the physical components that make up an Arduino project. At the heart of the Arduino ecosystem is the **Arduino board**, a microcontroller that acts as the brain of your project. There are several types of Arduino boards, but we'll start by focusing on the most popular ones:

- **Arduino Uno**: This is the most commonly used board and is ideal for beginners. It has 14 digital input/output pins, 6 analog inputs, a USB connection, and a power jack for external power sources.

- **Arduino Nano**: A smaller version of the Uno, the Nano is great for projects where size is a concern. It has similar functionality but is more compact.
- **Arduino Mega**: For more complex projects that require more pins and memory, the Mega is a great choice. It has 54 digital I/O pins and 16 analog inputs.

Each Arduino board comes with a set of pins that allow you to connect to external components like LEDs, motors, sensors, and displays. These components are what allow Arduino to interact with the physical world.

The Arduino Software (IDE)

To communicate with your Arduino board and program it, you'll need to use the **Arduino Integrated Development Environment (IDE)**. The Arduino IDE is free software that you can install on your computer (Windows, macOS, or Linux). It is where you'll write and upload your code to your Arduino board.

When you open the Arduino IDE, you'll see a simple editor with space for writing your code, as well as tools for uploading your code to your Arduino. The IDE is straightforward and designed to make programming as easy as possible for beginners.

Your First Steps with Arduino Programming

Now that you have a basic understanding of what Arduino is, let's take the first few steps in programming and creating with Arduino.

Step 1: Set Up Your Arduino Board

- Unbox your Arduino board and plug it into your computer using a USB cable. If you're using the Arduino Uno, connect it to the USB port on your laptop or desktop.

Step 2: Install the Arduino IDE

- Download and install the Arduino IDE from the official Arduino website. It's free to download and use.

Step 3: Write Your First Program: Blinking an LED One of the first projects that every Arduino beginner tackles is the **Blink** example, where we program the Arduino to turn an LED on and off at regular intervals.

To control an LED using an Arduino, you can follow these simple steps.

First, in the setup() function, you'll set pin 13 as an output pin. This is because the LED will be connected to this pin, and you need to tell the Arduino to send signals to it.

Next, in the loop() function, you'll alternate turning the LED on and off every second. To do this:

- Use digitalWrite(13, HIGH) to turn the LED on.
- Then, wait for one second with the delay(1000) function.
- After that, use digitalWrite(13, LOW) to turn the LED off.
- Wait for another second with delay(1000) before repeating the cycle.

This code will make the LED blink on and off every second continuously.

Step 4: Upload and Test

- After writing the code, hit the "Upload" button in the Arduino IDE. If everything is set up correctly, you should see the onboard LED on your Arduino board blink on and off every second.

Congratulations! You've just written your first Arduino program. As simple as this project is, it demonstrates how code can control physical hardware, which is the foundation of all Arduino projects.

Unique Hook: How Arduino Has Changed the World: Real-World Applications of a Small, Powerful Platform

Imagine this: A small device sitting on a workbench, no bigger than a credit card, connected to a series of wires, sensors, and motors. It's not just an unassuming circuit board—it's a tool that has the power to change lives. This is the magic of **Arduino**, a small microcontroller that has inspired innovators around the world to turn their ideas into tangible, real-world solutions. In fact, some of the most groundbreaking inventions in recent years, from life-saving medical devices to smart home technology, were powered by Arduino.

But what makes Arduino such a powerful tool? It's the combination of **affordability**, **simplicity**, and **accessibility**. It gives anyone, from students and hobbyists to professionals, the ability to create complex, interactive projects without needing to be an expert in electronics or coding. This small board has become the **foundation for some of the most innovative solutions in the world**—and it's something you can use to create your own life-changing inventions.

Real-World Applications: How Arduino Is Revolutionizing Everyday Life

Let's explore how Arduino has been used in real-world applications to **solve everyday problems, enhance lives**, and **push the boundaries of innovation**.

1. Life-Saving Medical Devices

One of the most significant applications of Arduino is in the medical field. Arduino has powered innovations that **save lives** and **improve patient care**. For example, Arduino is commonly used to create affordable **health-monitoring devices** for people in remote areas or for those with limited access to healthcare. A prime example is the development of **open-source ventilators**. During the COVID-19 pandemic, engineers and medical professionals around the world used Arduino to design low-cost ventilators to treat patients in under-resourced hospitals.

In addition to ventilators, **Arduino-based heart rate monitors**, **blood oxygen sensors**, and **glucose monitoring devices** have been created, allowing for real-time data collection that helps doctors and patients better manage chronic conditions. The affordability and flexibility of Arduino hardware have made it easier to build these devices at a fraction of the cost of commercially available models—making healthcare more accessible and cost-effective.

2. Environmental Conservation: Monitoring and Protecting Our Planet

Arduino has also played a critical role in **environmental conservation** by helping to monitor and protect our planet. With the increasing threat of climate change and environmental degradation, we need innovative tools to monitor and analyze environmental factors in real time. Arduino has been used to develop **automated weather stations**, **air quality sensors**, and **water quality monitoring systems**.

For example, **Arduino-based sensors** can be used to measure air pollution levels, track deforestation, or monitor water quality in lakes and rivers. These affordable, easily deployable systems make it possible to gather critical data on the environment, enabling better decision-making for governments, environmental organizations, and communities. **Citizen scientists** can even use Arduino to monitor their local environment, contributing to global conservation efforts.

3. Smart Home Automation

The future of our homes is becoming **smarter**, and Arduino is at the heart of this revolution. Arduino enables DIY enthusiasts to build customized **smart home automation systems** that are not only functional but also incredibly affordable. Whether it's controlling your lights, temperature, or security system from your phone, Arduino allows you to integrate and automate everyday tasks with ease.

For example, imagine a system that **automatically adjusts your thermostat** based on the weather or your daily schedule. Or a **smart irrigation system** that waters your garden only when the soil moisture level is low, conserving water while keeping your plants healthy. These aren't just futuristic ideas—they're everyday projects that can be built with Arduino.

Furthermore, **Arduino-based home security systems** with motion sensors, cameras, and smart alarms have become popular DIY projects. With the help of Arduino, people can design security systems that suit their specific needs, making their homes safer while cutting down on installation costs.

4. Robotics and Automation

Arduino has made robotics more accessible than ever before. Thanks to Arduino, anyone—from hobbyists to professional engineers—can build robots that can walk, talk, sense their environment, and perform complex tasks. Arduino has been used to power everything from **autonomous vehicles** to **robotic arms** and **swarming robots**.

One inspiring example is the **Arduino-powered robot assistants** used in hospitals and eldercare facilities. These robots are designed to help carry out repetitive tasks such as delivering supplies or assisting with physical therapy. By utilizing Arduino's versatility, engineers can create robots that can be easily programmed to perform a range of tasks, all without a huge budget.

Arduino is also a popular tool for teaching **robotics and programming**. With accessible learning resources and user-friendly software, students of all ages can build their own robots, giving them hands-on experience in the fields of engineering, coding, and problem-solving. Arduino has transformed robotics from a high-cost, high-tech field into one that is accessible to anyone with an interest in building and creating.

5. Accessibility and Assistive Technology

One of the most heartwarming and impactful uses of Arduino is in the field of **assistive technology**. Arduino has been used to design **affordable devices** that help people with disabilities live more independent lives. From **voice-activated systems** for people with limited mobility to **adaptive controllers** for gamers with physical disabilities, Arduino has been at the forefront of empowering people to overcome barriers.

For example, **Arduino-based prosthetic limbs** are being developed that can be customized and fitted for individuals at a fraction of the cost of traditional prosthetics. Arduino-powered systems that use sensors and motors can help individuals regain control of their environment and their daily activities, improving their quality of life.

6. The Maker Movement: Creativity, Innovation, and Community

The real power of Arduino lies not just in the products it enables, but in the **creative community** that has sprung up around it. The **Maker Movement**, driven by Arduino, encourages individuals to create, tinker, and share their projects with others. This community-driven approach has led to an explosion of **innovative DIY projects** and **inventions** that solve real-world problems.

From educational tools like **interactive learning kits** to **automated gardening systems** and **self-driving cars**, Arduino has fostered a spirit of creativity, collaboration, and innovation. With endless resources, forums, and tutorials, Arduino is a platform where anyone with an idea can turn it into reality—and inspire others to do the same.

Why Arduino? Why You?

You might be thinking: "These are incredible applications, but how can I get started with Arduino?" The truth is, Arduino is one of the most **accessible platforms** for beginners to learn programming and electronics. In this book, we'll guide you step by step through creating your own projects—whether it's a **smart weather station**, a **robot**, or even your own **life-saving device**.

The beauty of Arduino is that it gives you the **tools and knowledge** to take any idea, no matter how big or small, and turn it into a working, real-world application. The possibilities are endless, and it all starts with a single step—**learning to code and understand the hardware** that powers these innovations.

In the chapters ahead, we'll dive into the core concepts of Arduino programming and electronics, showing you how to build projects that can **solve problems, enhance daily life**, and ultimately, **make the world a better place**.

Focus on the Journey: From Beginner to Tech Innovator – Your Path to Building Functional Robots and Cutting-Edge Projects

Embarking on a journey to learn Arduino programming is not just about acquiring technical skills—it's about transforming yourself from someone who's new to coding and electronics into a creator of **functional, innovative tech projects**. This transformation is a process, a **step-by-step evolution** that begins with curiosity and ends with mastery. It's a journey of growth, problem-solving, and discovery, where every obstacle you overcome and every new skill you learn brings you closer to your ultimate goal: creating your own robots, smart devices, and tech projects that make a difference in the world.

But let's be clear from the start: **this journey isn't about perfection**—it's about progress. As you start your Arduino journey, you may encounter moments of frustration, confusion, or doubt. You might feel overwhelmed by the complexity of coding or the intricacies of electronics. But with each small step you take, you will gain confidence, build your skills, and witness the **evolution of your own abilities**. By the end of this journey, you will be amazed at the progress you've made, from having little to no understanding of microcontrollers to confidently creating your own robots and tech projects.

Your First Steps: Beginning the Journey with Curiosity

Every great journey begins with the decision to take that **first step**. When you first pick up an Arduino, you may have no prior experience in coding, electronics, or robotics. That's okay. **Everyone starts somewhere**, and even the most accomplished engineers, programmers, and makers once stood exactly where you are now.

The first chapters of this book are designed to ease you into the world of Arduino programming, breaking down complex concepts into manageable pieces. You'll start by learning the **fundamentals**: how to set up your Arduino board, how to write and upload simple code, and how to connect basic components like LEDs and sensors. These small successes will give you the confidence you need to push forward and tackle more challenging projects.

At this stage, your progress may feel incremental, and that's normal. Just remember, every time you succeed in completing a simple project—whether it's blinking an LED or turning on a motor—you're gaining a deeper understanding of how things work. You're learning the building blocks that will allow you to construct more **complex systems** in the future.

Leveling Up: Gaining Confidence and Skills

As you build your foundational knowledge and gain confidence with the basics of Arduino programming, you'll begin to see your potential expand. **New concepts** will start to make sense, and you'll be able to take on more complex projects. You might find yourself creating your first functional robot, writing more sophisticated code, or integrating sensors that allow your projects to interact with the world in meaningful ways.

At this stage, you'll start to realize something incredibly powerful: **the ability to make your ideas come to life**. As you add more components to your projects—whether that's motors, sensors, or actuators—you'll see your creations start to take shape. This could be as simple as a robot that can move in response to light or a weather station that reads real-time environmental data.

You may also begin to experiment with adding wireless communication or learning to debug your code. These challenges will require you to think critically, problem-solve, and even fail at times. And that's the beauty of the journey: each failure is a lesson in disguise, a crucial part of learning. The ability to **troubleshoot** and **improvise** is an essential skill that will serve you in future projects.

At this point, you might be amazed at how far you've come—from a beginner who struggled with setting up a simple circuit to someone who can now program a **functional device**. **This is the power of Arduino**: it teaches you to think like an engineer, to break down problems into manageable pieces, and to build solutions from the ground up. And remember, the more projects you complete, the more confidence you will gain in your own abilities.

Becoming a Creator: Building Real-World Robots and Tech Projects

By now, you're well into your journey, and you're not just programming basic circuits—you're now **building functional robots and tech projects**. Whether it's a **line-following robot**, an **automated plant-watering system**, or a **smart home device**, you are now bringing your ideas to life in the real world.

At this stage, you may feel a sense of pride as you see your projects working seamlessly, knowing that each successful build represents a significant milestone in your journey. This is where the real magic happens: you are not just learning; you are **creating**. And as you push the boundaries of what you can do, you'll start thinking about how you can improve your designs, **innovate**, and take on even more ambitious projects.

This stage is about **diversifying your skills**. You'll learn to integrate **advanced sensors**, explore **communication protocols** like Bluetooth and Wi-Fi, and dive into more **complex coding techniques**. Whether it's implementing **AI** algorithms for your robot, creating a fully autonomous system, or building a project that interacts with the internet (like a weather station with cloud integration), you will start to see the **infinite possibilities** for what you can create with Arduino.

Innovation and Mastery: Taking It to the Next Level

Once you've built several working robots and tech projects, you'll begin to notice a shift: you are no longer simply following tutorials or building pre-designed projects—you are **innovating**. You'll have the tools and knowledge to identify problems and create solutions of your own. The projects you build will no longer just be functional; they will be your own personal creations, reflecting your unique interests, ideas, and creativity.

At this stage, you may also find yourself **exploring new technologies** and incorporating them into your projects. You might combine Arduino with other platforms like Raspberry Pi, or integrate machine learning into your robot's decision-making process. You'll start to push the limits of what you thought was possible, and that's when things get truly exciting.

Your projects will start to evolve into something more than just DIY experiments—they'll become prototypes, models, and even real-world products. You might build an **autonomous drone**, a **robotic arm** for industrial use, or a **smart device** that can help solve a problem in your own life or community. Whatever you choose to build, you'll have the skills, knowledge, and confidence to **turn your ideas into reality**.

Envisioning the Future: What's Next in Your Journey

As you reflect on how far you've come—starting from a beginner with little to no experience, to someone capable of building and programming robots and tech projects—you'll realize that the journey doesn't end here. In fact, it's just the beginning. The skills you've gained with Arduino are **transferable** to many other fields in technology, from electronics engineering to artificial intelligence, IoT, and robotics.

The beauty of Arduino is that it's an ever-evolving field. As you continue your journey, new technologies, ideas, and opportunities will open up to you. You will never stop learning, experimenting, and creating. You'll continue to grow as an engineer, a programmer, and an innovator.

And as you continue to envision your future, know this: **you are not just a beginner anymore**. You are now capable of designing and building your own tech innovations. You've taken the first step toward becoming an engineer, a creator, and a problem solver. The future is yours to shape—and the journey you're on will help you do just that. So, let's move forward with confidence and excitement for what's to come. The world is waiting for your ideas.

Set Expectations: Understanding the Road Ahead in Arduino Programming

Before you embark on this journey of learning Arduino programming, it's essential to set the right expectations. While excitement and curiosity are great motivators, having a clear understanding of what you can realistically achieve and what challenges you might face will help you stay focused, motivated, and successful. Setting expectations not only prepares you mentally but also ensures you approach this journey with the right mindset, avoiding frustration when things don't go as planned.

1. Learning Takes Time and Patience

The first thing to understand is that **Arduino programming is a process**, and like any skill, it requires time and practice to master. You won't become an expert overnight, and that's okay. **Patience** is one of the most important traits to develop as you move through your learning. It's not uncommon to encounter obstacles or to feel overwhelmed by some of the more technical aspects of the projects you undertake.

Expect to make mistakes—and embrace them as opportunities to learn. In fact, many of the most successful engineers and programmers attribute their success to the lessons they've learned from failure. Debugging a project that doesn't work as expected, or realizing that your circuit isn't behaving as it should, is a natural part of the journey. **Mistakes are stepping stones**, and with each one, you'll gain the experience that moves you closer to mastery.

2. Building Blocks: Starting Small to Build Bigger Things

When you first begin with Arduino, it's important to understand that **small projects** lay the foundation for larger ones. The book will guide you through the basics: setting up your board, learning to code simple commands, and working with small components like LEDs, motors, and sensors. At first, the projects may seem simple, even trivial, but don't be deceived—the foundational knowledge you are gaining is crucial for more complex designs in the future.

Think of each small project as a **building block**. Each time you successfully complete a basic project—whether it's turning on an LED or making a sensor read data—you're strengthening your understanding of both hardware and software. These projects may feel small now, but they are the **cornerstones** that will support larger and more intricate systems. As you move forward, you'll start combining these smaller building blocks into **more sophisticated projects**. Gradually, you will build the skills needed to take on more complex creations, such as robots, IoT devices, or even smart homes.

3. Expect to Be Challenged—But Know You Can Overcome

Arduino programming can present challenges that may seem intimidating at first. From wiring circuits correctly to writing complex code, you may face moments of **frustration**.

You might even encounter roadblocks where things don't work as expected despite your best efforts. **These challenges are part of the learning process**, and they're an essential part of the growth that will lead to your success.

It's important to understand that **problem-solving is a skill** that takes time to develop. When something doesn't work, take a step back and **analyze the problem** logically. Break down the issue into smaller parts and **troubleshoot** systematically. This process will teach you to think like an engineer and a programmer, to approach problems with a calm and methodical mindset, and to develop critical problem-solving skills.

In the beginning, you may struggle with coding syntax, understanding logic flows, or integrating different components, but as you work through these challenges, your ability to resolve issues will improve dramatically. Soon, you'll begin to see problems as exciting puzzles to solve rather than insurmountable barriers.

4. Hands-On Learning: The Power of Experimentation

One of the unique aspects of Arduino programming is the **hands-on nature** of the learning process. Unlike purely theoretical subjects, Arduino allows you to immediately apply what you've learned to create something tangible. You'll spend a lot of time experimenting, and it's important to recognize that **trial and error** is an inherent part of the journey.

When building projects, you'll find that some components might not behave as you expect or your code might not produce the desired results. Don't be discouraged. The most valuable lessons often come from **hands-on experimentation**, and in the process, you'll become more comfortable with testing out ideas, learning from failures, and refining your solutions.

Experimentation is the fastest way to gain deep, lasting knowledge. You will learn best by **doing**—even if it means making mistakes along the way. Over time, you'll gain intuition about how components interact, how to structure your code, and how to troubleshoot issues effectively.

5. Real-World Applications: Start Simple, Dream Big

Arduino programming is not only about learning the technical skills of coding and electronics—it's also about applying those skills to create **real-world projects**. As you progress, you'll start imagining how you can use your newfound knowledge to build devices and robots that solve problems in the world around you.

It's crucial to understand that the projects you build at first may not seem groundbreaking, but they are valuable learning experiences. Perhaps you start by creating a simple temperature sensor, then move on to an automated plant-watering system. These small projects may seem simple, but they teach important concepts like **data collection**, **automation**, and **sensors** that are foundational for more complex systems.

You may find yourself dreaming of larger projects, like building an **autonomous robot**, designing a **smart home system**, or creating a **wearable device**. As you move through this book, you'll start to develop the skills necessary to turn those dreams into reality. But it's important to pace yourself. **Dream big, but start small.** The knowledge you gain from smaller projects will enable you to tackle more ambitious goals as you progress.

6. The Ongoing Journey: From Beginner to Expert

Lastly, setting expectations means understanding that **learning is a continuous process**. Arduino programming isn't a skill you simply "learn" and then move on from—it's a **journey of continuous growth**. Even as you become proficient in creating robots, automating systems, and programming devices, there will always be new challenges and technologies to explore.

As you dive deeper into more advanced topics like **machine learning**, **AI integration**, and **advanced electronics**, you'll continue to evolve as a maker. Even after you've completed the projects in this book, you'll find new areas of interest and opportunities to explore. The skills you acquire now will give you the foundation to continue building and innovating long after you've finished your first Arduino project.

The world of technology and innovation is constantly evolving, and as you build your projects, you'll be in a great position to continue learning, adapting, and contributing to the ever-expanding landscape of **coding**, **electronics**, and **robotics**.

Chapter 2: Understanding the Basics: What is Arduino and Why It's a Game-Changer

In the world of electronics and programming, **Arduino** has emerged as a transformative platform that empowers anyone—from complete beginners to seasoned engineers—to create innovative and functional projects. Whether you want to build a simple robot, automate your home, or create a smart device, Arduino offers a cost-effective and approachable way to get started. But what exactly is Arduino, and why has it become such a game-changer for makers, hobbyists, and professionals alike? This chapter will dive deep into the core principles of Arduino and explore why it's one of the most powerful tools available for both learning and creating.

1. What is Arduino?

At its core, **Arduino** is an open-source electronics platform that consists of both **hardware** and **software**. The hardware is typically a small microcontroller board that allows you to interact with the physical world by controlling sensors, lights, motors, and other electronic components. The software, called the **Arduino IDE (Integrated Development Environment)**, is where you write and upload the code that tells the hardware what to do.

The most common Arduino boards are equipped with a **microcontroller**—a small computer on a chip—that controls the entire board's functionality. These boards are designed to be simple and flexible, so anyone can use them, regardless of technical expertise.

Arduino can read inputs from a variety of sensors (like temperature, light, and motion sensors), process that data, and trigger outputs such as lights, motors, or displays. This interaction between software and hardware is what makes Arduino so powerful for building a wide range of projects.

2. The Open-Source Nature of Arduino: A Collaborative Revolution

One of the primary reasons for Arduino's widespread success is its **open-source** nature. Open-source means that both the hardware and software are free to use, modify, and distribute. This has fostered a massive, global community of creators who contribute to the platform by sharing designs, code, tutorials, and project ideas.

This open-source model has led to an ecosystem where people—from hobbyists to professional engineers—can collaborate, learn, and share ideas freely. Thousands of open-source libraries, tutorials, and example projects are available online, making it easier than ever to get started with Arduino, regardless of your experience level.

Because the code and designs are available for modification, you can make changes to suit your own needs, or even improve upon existing projects. This approach allows for **endless customization**, enabling makers to create unique devices that fit their specific goals.

3. A Platform for Learning: Accessibility and Empowerment

One of the most compelling reasons why Arduino is a game-changer is its **accessibility**. It's designed to be simple and easy to use, which makes it a great choice for beginners and educators. If you've never programmed or worked with electronics before, Arduino breaks down complex concepts into manageable chunks, allowing you to learn at your own pace.

The **Arduino IDE** provides an easy-to-understand programming environment with a vast library of built-in functions and examples. Writing code for Arduino is also beginner-friendly, using a language based on **C++** that is tailored to be easy for new users to grasp. What's more, Arduino's software allows for a smooth transition into more advanced programming languages as you grow in your knowledge.

Beyond programming, Arduino also introduces you to essential **electronics concepts**, such as wiring, power supply, sensors, and actuators. It's a hands-on learning experience where you can immediately apply what you've learned by building real-world projects. Instead of reading about concepts in a textbook, you can physically see and interact with the circuits you create.

4. Why Arduino Is a Game-Changer for Innovation

The true power of Arduino lies in its **ability to enable rapid prototyping** and **creative innovation**. Unlike traditional electronics and robotics platforms that may require complex setups, extensive equipment, and large budgets, Arduino provides an easy and affordable way to create sophisticated systems. Here's why it's a game-changer:

1. Affordable and Accessible

Arduino boards are inexpensive, with some starting as low as $10–$20. This affordability lowers the barrier for entry, making it possible for anyone to begin building their own tech projects. Whether you're a student, hobbyist, or entrepreneur, you don't need a large budget to get started with Arduino.

Additionally, the wide availability of components and accessories, like sensors, motors, displays, and shields (which are add-on boards), allows you to expand your projects without worrying about high costs. As the prices of components continue to drop, Arduino becomes even more accessible to a wider audience.

2. Ideal for Rapid Prototyping

In the tech and product development world, **rapid prototyping** is a crucial part of innovation. Arduino is a powerful tool for quickly turning ideas into functional prototypes. Whether you're designing a new gadget or testing a concept, Arduino allows you to test your ideas without spending months designing complex systems.

You can quickly create proof-of-concept models, test different configurations, and iterate on designs. This speeds up the process of turning an idea into a tangible product, which is why Arduino is favored by engineers, designers, and inventors alike.

3. Versatility Across Industries

Arduino isn't just for hobbyists—its versatility has made it popular in a wide range of industries. From **robotics** to **IoT**

(Internet of Things), **home automation**, and **art installations**, Arduino is used in various professional and commercial applications. Many professionals use Arduino to build prototypes for larger-scale projects, and some companies even use it in production systems.

Its compatibility with a wide range of components and technologies, such as **Bluetooth**, **Wi-Fi**, and **GPS**, makes it ideal for creating connected devices that integrate seamlessly with the world around us.

5. The Democratization of Technology

Arduino has played a crucial role in the **democratization of technology**. In the past, creating complex electronic devices required access to specialized knowledge and expensive equipment. However, Arduino's accessibility has made technology available to people from all walks of life, from teenagers building their first robot to seasoned engineers creating cutting-edge devices.

This democratization is fostering a global wave of **innovation**. It allows people who may not have had access to formal education in electronics and programming to still create real, functional devices. The creative freedom that Arduino offers has led to an explosion of ideas and projects, many of which have had real-world impact, from wearable health devices to smart home systems.

6. The Gateway to Robotics and IoT

Arduino is particularly popular in the fields of **robotics** and **Internet of Things (IoT)**, both of which are some of the most exciting technological frontiers today. As a beginner-friendly platform, it serves as a gateway for newcomers to start building robots and smart devices.

Robotics: By integrating sensors, motors, and controllers with Arduino, you can create robots capable of simple tasks or complex movements. Arduino enables you to program the robot to react to its environment, such as navigating obstacles, following lines, or even communicating with other devices.

Internet of Things (IoT): Arduino can also connect to the internet via modules like the ESP8266 or Ethernet shield, enabling you to create IoT devices that can send and receive data over the web. Whether it's a weather station, a smart light system, or a home automation hub, Arduino makes it possible to bring the Internet of Things to life.

7. Building a Community of Innovators

Another game-changing aspect of Arduino is the **community** that surrounds it. The Arduino community is a thriving global network of makers, hobbyists, educators, and professionals who share knowledge, projects, and ideas. Whether you need help troubleshooting your project, or you want to contribute your own inventions to the community, you'll find a wealth of support and inspiration in the Arduino ecosystem.

The open-source model fosters a spirit of **collaboration**, which enables continuous improvement and innovation. By working together, members of the Arduino community contribute to the platform's growth and help newcomers realize their creative potential.

Clear Explanation: How Arduino Fits into the Larger Landscape of Programming and Technology

Arduino is not just a standalone tool for learning programming and building basic projects; it is a fundamental building block in the larger landscape of technology, programming, and innovation. To understand how Arduino fits into this broader context, it's important to look at the intersection of hardware, software, and the global technological movement that is reshaping industries today.

1. Bridging the Gap Between Hardware and Software

At its core, Arduino acts as a bridge between **hardware** and **software**. Many traditional programming languages, like Python or Java, focus solely on software development—creating applications and programs that run on computers or mobile devices. However, Arduino takes things one step further by allowing you to program physical hardware. This integration between the digital (software) and physical (hardware) worlds is known as **embedded programming**, and Arduino provides an accessible and cost-effective way to dive into it.

Unlike traditional programming, which often deals with abstract concepts and distant outcomes, Arduino's focus on hardware allows for immediate, tangible results. You can see the impact of your code in real-time—whether it's turning on an LED, moving a motor, or responding to a sensor. This feedback loop reinforces your understanding of programming and its connection to the physical world, which is crucial for anyone looking to build or design real-world systems.

2. A Gateway to Internet of Things (IoT) and Smart Devices

Arduino plays a pivotal role in the **Internet of Things (IoT)**, which is one of the most exciting and rapidly growing sectors of technology. IoT refers to the growing network of interconnected devices—everything from home automation systems to wearable health trackers—that communicate and share data through the internet.

Arduino provides an ideal platform for **IoT experimentation** and prototyping. By connecting Arduino boards to sensors, actuators, and network modules like Wi-Fi or Bluetooth, users can create **smart devices** that interact with the world and communicate with each other. For example, an Arduino-based smart thermostat can measure temperature, adjust the room's heating or cooling, and send data to a smartphone app, all while being controlled remotely.

This ability to create smart devices is a stepping stone to the larger tech trends of **automation**, **artificial intelligence**, and **machine learning**, all of which are beginning to permeate nearly every industry.

3. Arduino in the Context of Modern Programming Languages and Platforms

Arduino is also part of a much larger **ecosystem** of programming languages, platforms, and frameworks. While Arduino uses a simplified version of **C++** for writing code, it introduces key programming concepts that are widely applicable across other platforms. As you progress, you can easily transition from Arduino to more advanced software development projects in **Python**, **Java**, or **C#**, learning valuable coding techniques and design patterns along the way.

Furthermore, Arduino integrates seamlessly with other tools and frameworks in the technology space. For example, it can work with **Raspberry Pi** (another popular platform for learning about electronics and programming), or with more complex platforms like **TensorFlow** and **OpenCV** for building intelligent, vision-based systems. This makes Arduino an important foundational piece for anyone looking to enter fields like **robotics**, **machine learning**, or **artificial intelligence**.

Comprehensive Overview: The History, Community, and Growing Industry of Open-Source Hardware

To truly understand the impact and significance of Arduino, it's crucial to explore its **history**, the **community** that has supported it, and the **open-source hardware** movement that it is part of. These elements work together to explain why Arduino has not only revolutionized personal electronics but also continues to drive innovation across industries.

1. The History of Arduino: From a Classroom Experiment to a Global Phenomenon

Arduino was founded in **2005** by a group of developers and engineers—**Massimo Banzi**, **David Cuartielles**, **Tom Igoe**, **Gianluca Martino**, and **David Mellis**—as a tool to help students, artists, and designers create interactive projects without needing a complex background in electronics. The team at Arduino wanted to develop an easy-to-use, affordable microcontroller platform that allowed users to explore physical computing.

Initially, the goal was to make a tool for students to experiment with interactive art and design projects. However, as the Arduino board gained attention and popularity, it rapidly became a versatile platform used by people across industries, from hobbyists and DIY enthusiasts to engineers and researchers. Today, Arduino has evolved into a global phenomenon, powering projects that range from simple LED circuits to sophisticated drones, smart homes, and even space exploration devices.

The name "Arduino" itself came from a bar in Ivrea, Italy, where the founders would often meet. Today, it is synonymous with hands-on learning, open-source innovation, and the democratization of technology.

2. The Open-Source Hardware Revolution

Arduino is part of the **open-source hardware** movement, a rapidly growing movement that allows people to share, modify, and build upon hardware designs and software code freely. Unlike traditional proprietary hardware platforms, where designs are locked behind patents and licensing fees, open-source hardware gives anyone the ability to study, improve, and redistribute designs.

The **open-source** model encourages **collaboration** and **innovation** in ways that were previously impossible. By making Arduino's designs publicly available, it enables anyone—whether an artist, student, entrepreneur, or engineer—to build upon existing projects, modify designs, and contribute new ideas back to the community. This shared approach fosters an **inclusive** environment where creativity thrives and innovation can happen at a rapid pace.

Arduino's open-source nature has contributed to a **global movement** in which individuals have the freedom to develop solutions to real-world problems, share those solutions with the community, and collectively advance the state of technology. For instance, medical devices, home automation systems, robotics, and renewable energy projects are all fields that have seen significant progress thanks to open-source platforms like Arduino.

3. The Arduino Community: Empowering a New Generation of Innovators

At the heart of Arduino's success is its **community**—a vibrant and diverse group of creators, educators, engineers, and enthusiasts who actively share knowledge and inspire others. The Arduino community consists of **forums**, **blogs**, **tutorials**, **project sharing websites**, and social media groups where users can learn from one another, collaborate on projects, and seek help when needed.

This global community of makers is driven by a shared passion for learning, innovation, and experimentation. With millions of people around the world working with Arduino, it has become one of the largest and most supportive communities of its kind. The wealth of online resources, including forums like the official **Arduino Forum**, **Instructables**, and **Hackster.io**, ensures that users can find solutions to almost any problem they encounter.

Moreover, the **Arduino Foundation** and various local groups host workshops, hackathons, and **maker fairs**—events designed to engage the community and showcase the incredible things that can be built with Arduino. These events help to bridge the gap between learning and real-world application, allowing participants to network, learn new skills, and get inspired by others.

4. Arduino in the Growing Industry of Open-Source Hardware and Technology

The rise of Arduino has been part of a larger trend toward **open-source hardware** and **DIY electronics** that has gained significant momentum in recent years. As traditional technology giants move toward **closed ecosystems** and proprietary platforms, open-source hardware is fostering **democratized innovation**. Arduino has inspired a number of similar platforms, such as **Raspberry Pi**, **BeagleBone**, and **Particle**, which cater to different audiences and applications.

The open-source hardware movement has not only led to the growth of a thriving DIY culture but has also spurred new industries. **Educational technology** (EdTech), for example, has been transformed by Arduino, with schools and universities incorporating it into their curricula to teach programming, electronics, and problem-solving skills in a hands-on, interactive way.

Moreover, many industries now leverage Arduino for **rapid prototyping** and product development. In fields such as **robotics**, **automation**, **smart cities**, **renewable energy**, and **healthcare**, Arduino enables startups and large corporations alike to test new ideas, create proof-of-concept models, and build solutions without the need for large-scale infrastructure or high budgets.

Interactive Exercise: Setting Up Your Arduino IDE (Integrated Development Environment) and Seeing Immediate Results

One of the best ways to start learning Arduino programming is to dive right in. In this interactive exercise, you'll set up your Arduino IDE (Integrated Development Environment), which is the software tool that allows you to write, compile, and upload code to your Arduino board. This process is easy, fun, and hands-on—perfect for making sure you see immediate results and feel accomplished as you begin your journey into coding and electronics.

The Arduino IDE is user-friendly, making it accessible even if you have no prior experience with programming or electronics. By the end of this exercise, you'll have the confidence to write your first Arduino program, upload it to your board, and see the results of your efforts in real time. Let's break down the steps you'll take to get your IDE set up and your Arduino board working with your first simple project!

Step 1: Downloading and Installing the Arduino IDE

Before you can start programming, you'll need to download and install the Arduino IDE. It's available for Windows, macOS, and Linux, so no matter your operating system, you'll be able to follow along.

1. **Visit the Arduino website**: Go to www.arduino.cc to download the latest version of the IDE.
2. **Choose the right version**: Select the version that corresponds to your operating system (Windows, macOS, or Linux).

3. **Install the IDE**: Once the file is downloaded, open the installer and follow the on-screen instructions. For most users, the default settings will work just fine.
4. **Launch the IDE**: After installation, open the Arduino IDE from your desktop or applications folder.

Step 2: Connecting Your Arduino Board

Once you've got the IDE installed, it's time to connect your Arduino board to your computer. If you're using an **Arduino Uno** or **Arduino Nano**, this will be simple to do:

1. **Plug your Arduino board into your computer**: Use the USB cable that came with your board to connect it to an available USB port on your computer.
2. **Select the right board in the IDE**:
    - In the Arduino IDE, click on the **Tools** menu at the top of the screen.
    - Select **Board** and choose the model of your Arduino board (e.g., Arduino Uno, Arduino Nano, etc.).
3. **Select the correct port**:
    - In the **Tools** menu, go to **Port** and select the correct serial port to which your Arduino is connected. If you're unsure, try unplugging your Arduino and seeing which port disappears from the list, then reconnect your Arduino and select that port.

Step 3: Writing Your First Program (The Blink Sketch)

Now that your Arduino is connected and the IDE is configured, let's write a simple program called a **sketch**. The first program every Arduino user writes is typically the **Blink**

Sketch, which simply turns an onboard LED light on and off at regular intervals. This basic example will give you immediate, tangible results and show you how to interact with your hardware.

1. **Open the Blink Sketch**:
     - In the Arduino IDE, go to **File** > **Examples** > **01.Basics** > **Blink**. This will open a new window with the code that makes an LED blink.
2. **Understand the Code**:
     - **void setup()**: This function runs once when the program starts. It sets up the necessary configurations, such as telling the Arduino to use a certain pin for output (in this case, pin 13, which is usually connected to the onboard LED).
     - **void loop()**: This function runs repeatedly after setup() is completed. The LED will blink on and off because of the digitalWrite() commands within the loop, which turn the LED on and off, and the delay() function, which pauses the program for a certain number of milliseconds.

Here's the basic code for the Blink Sketch in Arduino:

In the setup() function, you're setting pin 13 as an output pin, so the Arduino knows that it will send a signal to control the LED connected to that pin.

In the loop() function, the program continuously alternates between turning the LED on and off:

- The digitalWrite(13, HIGH) command turns the LED on.

- The delay(1000) function pauses the program for 1 second (1000 milliseconds).
- The digitalWrite(13, LOW) command turns the LED off.
- The delay(1000) function pauses again for 1 second.

This cycle repeats endlessly, causing the LED to blink on and off every second.

3. **Upload the Code**:
    - Once you understand the code, it's time to upload it to your Arduino. Click on the **Upload** button (the right arrow) in the top left corner of the IDE. The IDE will compile the code and send it to your Arduino board.
    - Once the upload is successful, you should see the LED on your Arduino board blinking on and off every second.

Step 4: Seeing the Results and Reflecting on Your Progress

Now, you've completed your first project! The LED is blinking, and you've just programmed your Arduino board to do something interactive. This may seem like a simple exercise, but it's the perfect introduction to the world of Arduino programming and hardware interaction.

What You've Learned So Far:

- **Setting up the Arduino IDE**: You've installed the software that you'll be using throughout your Arduino projects.

- **Connecting your Arduino board**: You've learned how to connect your Arduino to your computer and select the correct settings in the IDE.
- **Writing a basic program (sketch)**: You've written your first Arduino program, called the **Blink Sketch**, and learned how to control hardware (the onboard LED) using simple code.
- **Uploading and testing your code**: You've successfully uploaded the program to your Arduino board and watched as the LED blinked, showing the immediate results of your work.

Step 5: Experiment and Modify the Code

The next step in this interactive exercise is to encourage you to **experiment** with the code you just wrote. Here are some ideas for how you can modify the program and observe different results:

1. **Change the Blink Rate**:
    - Modify the delay times to make the LED blink faster or slower. Try changing the values in the delay() function to something like delay(500) for half a second or delay(2000) for two seconds.
2. **Use a Different Pin**:
    - If you have an external LED or any other component you want to control, you can use a different pin on your Arduino board (e.g., pin 9, pin 8) and connect your LED to that pin. Don't forget to update the pin number in the code.
3. **Add More LEDs**:
    - If you have multiple LEDs, try adding them to different pins and writing code to make them blink in a sequence or pattern.

The Takeaway:

This exercise gives you a great introduction to the Arduino platform and programming basics. You've already completed your first project, and the best part is that this is only the beginning! As you work through the exercises in this book, you will keep building on this knowledge to create more complex and exciting projects.

By taking immediate action and seeing results from the very first exercise, you've laid a solid foundation to become proficient in Arduino programming. Every project you build from here will help you gain more confidence and skills as you work your way through the world of coding, robotics, and smart tech!

Chapter 3: Your First Project – Blinking an LED and Understanding Code

In this chapter, we'll build on the foundational knowledge you gained from setting up your Arduino IDE and connecting your board. The project we'll work on is both simple and essential: **Blinking an LED**. This classic "first project" not only demonstrates how to interact with your Arduino hardware, but it also introduces you to the essential concepts of coding that will serve as the building blocks for all future projects.

You'll learn how to control an LED, understand the structure of your first Arduino program (also called a **sketch**), and break down the code step-by-step. By the end of this chapter, you'll have a solid understanding of how to manipulate hardware using code, and you'll have gained confidence in writing and uploading your first program to your Arduino.

The Project: Blinking an LED

The goal of this project is to get an LED to blink on and off at regular intervals. LEDs are a simple but powerful way to interact with electronic components, and blinking one is an easy yet effective way to understand the basic programming structures in Arduino.

To start, you'll use the **onboard LED** that's already built into most Arduino boards. On an **Arduino Uno**, this LED is connected to **pin 13** by default, so you don't need any additional components to get started.

If you're working with a different Arduino model, don't worry—the process will be similar. If you want to use an **external LED**, we'll cover how to connect it later in the chapter.

Step 1: Understanding the Code

Before we dive into the code, let's explore its structure so you understand exactly what's happening. The **Blink Sketch** is a basic program that uses two main sections: **setup()** and **loop()**.

Here's the code we'll be working with:

In the setup() function, we set pin 13 as an output, so the Arduino knows it will be used to control the LED.

In the loop() function, the program continuously does the following:

1. Turns the LED on by using digitalWrite(13, HIGH).
2. Waits for 1 second with delay(1000).
3. Turns the LED off using digitalWrite(13, LOW).
4. Waits another second with delay(1000) before repeating the cycle.

This creates a blinking effect for the LED, turning it on and off every second.

Breaking Down the Code

1. setup() **Function**:
    - **Purpose**: The setup() function runs once when the program starts. It's where you'll define the

initial settings for your project, such as configuring the pins on the Arduino board.

Code Explanation:

The line:

cpp

Copy code

pinMode(13, OUTPUT);

This line tells the Arduino that **pin 13** will be used as an **output pin**.

In simpler terms, you're informing the Arduino that you want to control the LED (which is physically connected to pin 13) using code. By setting the pin to **OUTPUT**, you enable the Arduino to send signals to it.

When a pin is set to **OUTPUT**, you can control it by sending signals like **HIGH** (which powers the LED on) or **LOW** (which turns the LED off). This is how you can control not just LEDs, but also motors, relays, or other components connected to the Arduino.

2. **loop() Function**:
   - **Purpose**: The loop() function runs continuously after the setup() function finishes. This is the heart of your program, where all repeated actions happen.

The code loops through these actions repeatedly, creating a flashing effect where the LED turns on, waits for one second, turns off, and waits again.

Step 2: Uploading the Code

Now that you understand the structure of the code, it's time to upload it to your Arduino board and watch it come to life. Here's how you can upload the code:

1. **Open the Arduino IDE**: If you haven't already, launch the Arduino IDE that you installed earlier.
2. **Copy the Code**: Copy the code from the above section and paste it into a new sketch in the Arduino IDE.
3. **Select the Right Board and Port**:
    - In the Arduino IDE, go to the **Tools** menu.
    - Select **Board** and choose the model of your Arduino (e.g., Arduino Uno).
    - Then, select **Port** and choose the correct port to which your Arduino is connected.
4. **Upload the Code**: Click the **Upload** button in the Arduino IDE (the right arrow). The IDE will compile the code, upload it to your Arduino board, and you should see the onboard LED begin to blink after a few seconds.

Step 3: Modifying the Code and Experimenting

Once the LED is blinking, you have a perfect opportunity to experiment with the code and make it your own. This is one of the most important aspects of learning Arduino: **tinkering** with the code and seeing how changes affect the results. Here are a few things you can try:

1. **Change the Blink Rate**:
    - Modify the delay values to make the LED blink faster or slower. For example, try changing the delay from 1000 to 500 for a faster blink, or to 2000 for a slower blink.
2. **Add More LEDs**:
    - If you have an external LED, you can add it to your circuit and modify the code to control both LEDs. Let's say you connected an external LED to pin 8. You would add the following to the setup() function:
3. **Experiment with More Delays**:
    - Try using different delay values for each LED. You can make one LED blink faster than the other, creating a staggered effect.
4. **Use Different Pin Modes**:
    - Instead of pinMode(13, OUTPUT), try using different pins (e.g., pin 8, pin 9, etc.) to control additional LEDs or components.

Step 4: Understanding the Concepts Behind the Code

Let's take a moment to reflect on the key concepts this project teaches:

- **Output Pins**: You've learned how to configure pins as output in the setup() function using pinMode(). This is crucial for controlling components like LEDs, motors, and more.
- **Digital Signals**: By using digitalWrite(), you've learned how to send digital signals (either HIGH or LOW) to pins. This is how you turn things on and off in your projects.

- **Delays**: The delay() function introduces pauses into the code. This is useful for controlling timing in your projects, whether it's blinking an LED or creating delays between sensor readings.
- **Code Structure**: You've now seen how the setup() function runs once, and the loop() function runs repeatedly, providing a framework for all your Arduino programs.

Step 5: Moving Forward

Now that you've completed your first project and gained an understanding of the code behind the LED blink, you're ready to take your Arduino skills to the next level. The best way to continue learning is through hands-on practice and experimentation. In the next chapters, we'll introduce more complex components and coding concepts that will expand your abilities and empower you to create more intricate projects.

But for now, take a moment to feel proud of your achievement! You've just written your first Arduino program and brought it to life. This is the beginning of your journey toward becoming proficient in Arduino programming and making your own inventions.

Simplifying the Process: A Step-by-Step Guide to Your First "Hello World" Project in Arduino

One of the most rewarding aspects of learning Arduino programming is seeing your code come to life. To make that experience as smooth as possible, we'll guide you through your very first "Hello World" project. For Arduino, this classic introduction to programming takes the form of a simple **LED**

Blink. It's the perfect beginner's project because it teaches you how to interact with both the software and hardware components of your Arduino board.

Why "Hello World"?

In programming, **"Hello World"** refers to the simplest program you can create to test if your environment is set up correctly and to get you started on the path to learning how code can interact with the world around you. For Arduino, the "Hello World" project isn't about displaying text—it's about blinking an LED, a small light on your board or an external component. Blinking an LED serves as a perfect introduction because it covers basic concepts such as:

- **Setting up your Arduino IDE** for writing and uploading code.
- **Working with basic Arduino syntax**.
- **Understanding how to use digital pins** to control hardware.
- **Using functions** like digitalWrite() and delay().

By the end of this chapter, you'll understand how to program Arduino to control hardware in a simple, intuitive way—and you'll feel accomplished after seeing your first project come to life.

The Components You'll Need:

Before we begin, let's list out the components required for this project:

1. **Arduino Board** (Uno, Nano, Mega, or any compatible board)

2. **LED** (Light Emitting Diode)
3. **220-ohm Resistor** (if you're using an external LED)
4. **Jumper Wires**
5. **Breadboard** (optional, for external LED setup)

If you're using the **onboard LED** (located on most Arduino boards), you won't need any additional components because it's already wired for you, usually connected to **pin 13**.

If you prefer to use an **external LED**, you will need to wire it up properly with a **220-ohm resistor** to prevent damaging the LED or the Arduino.

Step 1: Set Up Your Arduino IDE

The first step is to ensure your **Arduino IDE** is properly set up on your computer. If you haven't done so already, follow these steps to get started:

1. **Download the Arduino IDE**: Visit the official Arduino website (https://www.arduino.cc/en/software) to download the latest version of the Arduino IDE for your operating system (Windows, macOS, or Linux).
2. **Install the IDE**: Once downloaded, run the installer and follow the on-screen instructions. After installation, open the Arduino IDE.
3. **Connect Your Arduino Board**: Plug your Arduino into your computer using a **USB cable**. Ensure that the board is properly recognized.
4. **Select Your Board and Port**:
    - In the **Tools** menu, go to **Board** and select your Arduino model (e.g., Arduino Uno).

- Next, in the **Tools** menu, select **Port** and choose the correct port that corresponds to your Arduino board.

Step 2: The Basic LED Blink Code

Now that everything is set up, it's time to write the code for your first project. Follow these steps carefully to create a simple program that blinks an LED:

1. **Create a New Sketch**: In the Arduino IDE, click on **File > New** to create a new sketch.
2. **Enter the Code**: Copy and paste the following code into your sketch:

Step 4: Uploading the Code to Your Arduino

Once you've written the code, the next step is to upload it to your Arduino board. Here's how to do it:

1. **Connect the Arduino to Your Computer**: Ensure your Arduino board is connected to your computer via the USB cable.
2. **Upload the Code**: In the Arduino IDE, click the **Upload** button (the right arrow) located at the top-left of the window. The IDE will compile your code, upload it to the board, and you should see the onboard LED blink.

Step 5: Troubleshooting

If your LED doesn't blink right away, don't worry! Troubleshooting is part of the learning process. Here are some common issues you might encounter:

- **No Blinking LED**:
    - Ensure your Arduino is properly connected to your computer and recognized in the **Tools > Port** menu.
    - Double-check your code to ensure that **pin 13** (or the correct pin for your external LED) is selected.
- **External LED not working**:
    - If you're using an external LED, ensure the polarity is correct. The longer leg of the LED is the **positive (anode)** side, which should be connected to the digital pin (e.g., pin 13), and the shorter leg (cathode) should go to **ground** via the resistor.
- **Code Error Messages**:
    - If the Arduino IDE displays an error message when you try to upload, check the message for clues. It may be a syntax error (e.g., a missing semicolon) or a problem with the board/port selection.

Step 6: Experimenting with Variations

Once you see your LED blink, it's time to experiment and make the project your own! Here are a few ideas to modify the blink rate or add complexity:

- **Change the Blink Speed**: Adjust the delay times to make the LED blink faster or slower. Try delay(500) for a half-second blink or delay(2000) for a two-second delay.
- **Control Multiple LEDs**: If you've added more LEDs, you can control them in the same way. Just add another pinMode() for each new pin in the setup()

function and control each LED with its own digitalWrite() in the loop() function.
- **Fade the LED**: To make the LED gradually get brighter or dimmer, you can use the analogWrite() function (if your board supports **PWM** pins) instead of digitalWrite(). This allows you to control the brightness of the LED.

Step-by-Step Approach: Understanding the Code Structure, Syntax, and Logic

When starting with Arduino programming, it's essential not just to **copy and paste** code but to truly understand the **structure**, **syntax**, and **logic** behind every line. This foundational understanding will make you a more confident and effective programmer in the long run, as you'll be able to troubleshoot, modify, and even create your own projects from scratch.

Let's break down the Arduino code structure and syntax, focusing on each component's purpose.

1. Arduino Sketch Structure

Every Arduino program, also called a **sketch**, follows a similar structure. While the code itself can be as complex as the project demands, all sketches share a basic layout:

- **Setup Function (setup())**: This function runs once when the Arduino starts up, making it ideal for initializations such as setting pin modes and starting serial communication. It's where you tell the board what it needs to know before it starts performing repetitive tasks.

- **Loop Function (loop())**: This function is the heart of your program. After the setup() function runs, Arduino continuously loops through the code in loop(), repeating the same instructions indefinitely. This is where you place the actions you want the Arduino to perform over and over again, like blinking an LED or reading sensor values.

2. Arduino Syntax

In addition to the basic structure, understanding the syntax of Arduino code is critical for ensuring your program works as expected.

- **Variable Declaration**: Variables store values used by the program. For example, in the blinking LED example, the ledPin variable stores the pin number where the LED is connected. Here's how you declare it:
- **Comments**: Comments are essential for documenting code. Anything following // is a comment and won't affect the program. It's a way to explain your code to yourself or others reading it.
    - **Multiline Comments**: If you need to add more detailed explanations, you can use a block comment, like this:
- **Functions**: Functions like pinMode(), digitalWrite(), and delay() are pre-defined in Arduino libraries, and you use them to instruct the board to do specific tasks. Learning how to use functions properly is vital for writing clean, readable, and efficient code.

3. The Logic Behind the Code

Let's focus on the **logic** behind the blinking LED example. The purpose of the program is simple: turn the LED on for 1 second, turn it off for 1 second, and repeat. Here's how the logic flows:

1. **Setup the Environment**: In setup(), you configure the pin as an output. This step is crucial because it tells Arduino that pin 13 is going to send signals to an external device (in this case, the LED).
2. **Loop to Control the LED**: The loop() function continuously sends **HIGH** and **LOW** signals to the LED. These signals turn the LED on and off. The delay() function creates a pause between these actions, giving you a visible blink.

This simple logic forms the basis for many more advanced projects, and once you understand it, you'll be able to apply it to various devices and sensors.

Error Troubleshooting: Mini-Section for Beginners

Even experienced programmers encounter errors, and as a beginner, it's normal to face challenges. **Error troubleshooting** is a valuable skill that will help you build confidence early in your Arduino journey.

Here are some common beginner errors and how to fix them:

1. Compilation Errors

When you click "Upload" in the Arduino IDE, the program compiles the code before sending it to the Arduino board. If you get an error message, it usually means there's something wrong with the syntax or structure of your code.

Error Message: void setup() was not declared in this scope

- **Cause**: The setup() function is either missing or has a typo.
- **Fix**: Make sure your setup() function is written correctly and placed before loop().

2. Pin Configuration Errors

If your LED doesn't blink, it might be because of an incorrect **pin configuration**.

- **Error**: LED not turning on or off.
    - **Cause**: The pin number specified for the LED might not match the physical pin the LED is connected to.
    - **Fix**: Double-check the **pin number** in your code and ensure it matches the physical connection of your LED to the Arduino board. For example, if you're using pin 13, make sure the LED is connected to that pin.

3. Wiring Errors (for External LED)

If you're using an external LED, incorrect wiring is a common issue.

- **Error**: LED doesn't light up at all.
    - **Cause**: You may have connected the LED backwards, or the resistor value might be incorrect.

- **Fix**: Ensure the **longer leg (anode)** of the LED is connected to the Arduino's digital pin (e.g., pin 13), and the **shorter leg (cathode)** is connected to **ground (GND)**. Also, use a **220-ohm resistor** to limit the current flowing through the LED.

4. Power Issues

If the Arduino board isn't receiving power, it won't run your code.

- **Error**: Arduino not responding or not showing up in the IDE.
    - **Cause**: The board might not be connected to the computer properly, or the USB cable might be faulty.
    - **Fix**: Ensure the Arduino is correctly connected to the computer via a USB cable. Try using a different cable if necessary.

5. Common Debugging Tips

- **Use Serial Monitor**: To see real-time values from your code, use the **Serial Monitor** in the Arduino IDE. Add Serial.begin(9600); in setup() to begin serial communication, and then use Serial.println(variableName); in loop() to print out variable values.
- **Read the Error Messages Carefully**: The Arduino IDE often provides helpful hints. If you get an error message, read it carefully to locate the problem.

Chapter 4: Exploring Sensors and Actuators

Making Arduino Come to Life

In the previous chapters, you've learned the fundamentals of Arduino programming, from setting up your environment to writing simple code that blinks an LED. But Arduino's true power comes to life when it interacts with the world around it—**sensors** and **actuators** are the bridge that connects the digital world of your code with the physical world of real-time inputs and outputs.

In this chapter, we will explore how to integrate sensors and actuators with Arduino to create meaningful projects that respond to environmental changes, control devices, and automate tasks. Whether you want to build a weather station, a motion-activated light, or a simple robotic system, understanding how to use these components is essential for creating dynamic and interactive projects.

What Are Sensors and Actuators?

1. Sensors: The Eyes and Ears of Arduino

Sensors are electronic devices that detect physical phenomena, such as light, sound, motion, temperature, or humidity, and convert them into electrical signals that Arduino can read and process. They are the input devices that allow Arduino to sense the world around it.

Some common types of sensors include:

- **Temperature Sensors**: Detect ambient temperature and send it to the Arduino. Common examples

include the **LM35** and the **DHT11** (which also measures humidity).
- **Light Sensors**: Measure the amount of light in the environment, usually using a **photoresistor (LDR)** or **photodiode**.
- **Motion Sensors**: Detect movement in the environment. A **PIR (Passive Infrared)** sensor is commonly used for this purpose in motion-activated systems.
- **Proximity Sensors**: Detect the presence of nearby objects without physical contact, often using **ultrasonic** or **infrared sensors**.
- **Pressure Sensors**: Measure force or pressure, typically using **force-sensitive resistors (FSRs)** or **barometric pressure sensors** like the **BMP180**.

By integrating sensors with Arduino, you enable it to monitor conditions in the physical world and respond accordingly—whether that's adjusting the brightness of an LED, sending data to a cloud service, or triggering an alarm.

2. Actuators: Bringing Arduino to Life

Actuators are devices that convert electrical signals from Arduino into physical actions. They perform tasks or interact with the environment, completing the feedback loop of a sensor-actuator system.

Examples of actuators include:

- **Motors**: Convert electrical energy into mechanical motion. Motors can rotate wheels, move arms, or drive other mechanical systems. Common types are **DC motors**, **servo motors**, and **stepper motors**.

- **Relays**: Act as switches that can control high-power devices like lamps, fans, and motors, all controlled by low-power signals from Arduino.
- **LEDs**: Light-emitting diodes that display visual feedback or provide light for projects like traffic lights, status indicators, or ambient lighting.
- **Buzzers and Speakers**: Produce sound or beeps. These can be used for alerts, alarms, or even simple music generation in your projects.
- **Servos**: Special motors that allow for precise control of angles. They're commonly used for robotic arms or pan-tilt camera mounts.

Together, sensors and actuators enable Arduino to become more than just a microcontroller—they turn it into an intelligent system capable of interacting with the world around it.

Integrating Sensors and Actuators into Your Arduino Projects

Now that we know what sensors and actuators are, let's explore how to integrate them into a basic Arduino project. To make this process clear, we'll cover a simple **temperature-based fan control system**, where the system turns on a fan (an actuator) if the temperature (measured by a sensor) exceeds a threshold.

1. Gathering Your Components

For this project, you will need the following components:

- **Arduino Uno** (or any other Arduino board)
- **DHT11** temperature and humidity sensor

- **Relay module**
- **DC fan** (or any other low-power actuator)
- **Breadboard and jumper wires**
- **Resistor (10kΩ)** for the DHT11 sensor (optional depending on sensor model)

2. Wiring the Temperature Sensor

- **DHT11 Sensor**: This sensor has three pins: **VCC**, **GND**, and **DATA**.
    - Connect the **VCC** pin to **5V** on the Arduino.
    - Connect the **GND** pin to **GND** on the Arduino.
    - Connect the **DATA** pin to a **digital pin** (e.g., pin 7) on the Arduino.

3. Connecting the Relay and Fan

- **Relay Module**: The relay controls the high-voltage fan using a low-voltage signal from the Arduino. The relay has four pins: **VCC, GND, IN,** and **COM/NO**.
    - Connect **VCC** to **5V** on the Arduino.
    - Connect **GND** to **GND** on the Arduino.
    - Connect **IN** to any **digital pin** (e.g., pin 8) on the Arduino.
    - Connect the **COM** pin to the **positive terminal** of the fan.
    - Connect the **NO** (normally open) pin to the **positive terminal of the power source** for the fan.
    - Connect the **negative terminal** of the fan to **GND**.

4. Writing the Code

The goal is for the fan to turn on when the temperature goes above a certain threshold (e.g., 30°C). Here's how to approach the code:

1. **Read Data from the Sensor**: You'll use the **DHT** library to read the temperature and humidity from the DHT11 sensor.
2. **Control the Relay Based on Temperature**: If the temperature exceeds a set value, the relay will be triggered to turn on the fan.
3. **Display Feedback**: You can use the **Serial Monitor** to display the current temperature and the status of the fan.

5. How the Code Works

- The **DHT11 sensor** reads the temperature.
- The Arduino reads this temperature data and compares it to the threshold (30°C).
- If the temperature exceeds 30°C, the relay is activated via the **RELAY_PIN**, turning on the fan.
- The **Serial Monitor** outputs the current temperature and the status of the fan to keep you informed of the system's behavior.

Common Challenges and How to Solve Them
1. Sensor Not Reading Correctly

- Ensure the wiring is correct and that the **DHT11** sensor is connected to the right pin.
- If the readings are inaccurate, try recalibrating or replacing the sensor. The **DHT11** has limitations and

is not as accurate as higher-end sensors like the **DHT22**.

2. Fan Not Turning On

- Double-check the relay wiring. Make sure the **COM** and **NO** pins are connected correctly to the power source and the fan.
- Ensure that the relay module is rated for the current required by your fan.

3. Code Not Working

- Make sure you've installed the necessary libraries for the DHT11 sensor. You can find and install them from the Arduino Library Manager.
- Verify that all pins in the code match the actual pins you are using in your circuit.

Focus on Sensory Interaction: Making Projects More Dynamic with Sensors

One of the most exciting aspects of working with Arduino is the ability to interact with the physical world in a meaningful way. While early projects like blinking LEDs or controlling motors provide a solid foundation, the true power of Arduino lies in its ability to sense and respond to changes in the environment around it. Sensors are the key to this interaction, transforming static, programmed responses into dynamic, real-time systems that can react to temperature, light, motion, and more.

In this section, we'll explore how **sensors**—such as temperature, light, and motion sensors—can be used to create interactive, responsive projects. These sensors serve as the "eyes," "ears," and "feelers" of your Arduino system, allowing it to sense external conditions and adjust its behavior accordingly. Whether you're building a smart home system, a security device, or a robot that can react to its surroundings, sensors are essential for adding **dynamic capabilities** to your Arduino projects.

What Makes Sensors So Powerful?

Sensors allow Arduino to detect **real-world changes** and act based on that information, essentially turning your project into an interactive system rather than a simple, static one. By incorporating sensors, you can give your projects the ability to:

- **Respond to environmental changes**: A sensor can detect changes in the environment, such as an increase in temperature or a change in light levels, and trigger a corresponding response in the system, like activating a fan or turning on a light.
- **Automate tasks**: Instead of needing manual input, your system can automatically adjust its behavior based on what it senses. For example, an automatic lighting system can turn on a light when it detects a person entering the room, or a motion-activated security system can trigger an alarm when it senses movement.
- **Enhance interaction**: Sensors enable your project to interact with its surroundings, making it more engaging and user-friendly. Imagine a robot that adjusts its speed when it detects obstacles, or a smart garden that waters plants based on the moisture level of the soil.

In short, sensors are the gateway to creating **intelligent, autonomous systems** that can sense and react to the world in real-time.

Types of Sensors for Dynamic Interaction

Now let's dive into a few of the most commonly used sensors in Arduino projects: temperature, light, and motion sensors. These sensors are often used in a wide range of applications and serve as excellent starting points for creating dynamic and interactive systems.

1. Temperature Sensors

Temperature sensors are a fundamental tool for adding environmental awareness to your projects. These sensors measure the ambient temperature and send that data to the Arduino for processing. One of the most commonly used temperature sensors in Arduino projects is the **DHT11** or **DHT22** (for more accurate readings).

Applications of temperature sensors:

- **Climate control systems**: Automatically adjust the settings of a heating or cooling system based on the room's temperature.
- **Smart appliances**: Turn a fan or air conditioner on or off based on the temperature detected in a room.
- **Weather stations**: Track the temperature in real-time and log the data for further analysis.

Example project: You could create an automatic fan system that turns on when the temperature exceeds a certain threshold. By programming the Arduino to monitor the temperature and activate an actuator (like a fan) when necessary, your system would dynamically respond to environmental changes.

2. Light Sensors

Light sensors, such as the **photoresistor (LDR)**, detect the intensity of light in their surroundings. When the light level changes, the resistance of the sensor changes, which can be read by the Arduino to trigger an action.

Applications of light sensors:

- **Automatic lighting systems**: Turn lights on when it gets dark or off when it gets light.
- **Solar-powered systems**: Charge batteries or power devices based on available sunlight.
- **Security systems**: Use light sensors to detect when lights are turned on or off in a room, which can indicate movement or activity.

Example project: Build an **automatic night light** that turns on when it gets dark outside. Using a photoresistor and Arduino, you can program the system to activate the light as soon as it detects a drop in ambient light levels, creating a dynamic and energy-efficient lighting solution.

3. Motion Sensors

Motion sensors are one of the most commonly used types of sensors in Arduino projects. A **PIR (Passive Infrared)** sensor detects changes in infrared radiation, typically caused by the movement of warm objects (like humans or animals). This makes them ideal for applications where you want to detect the presence or movement of people.

Applications of motion sensors:

- **Security systems**: Trigger alarms, cameras, or lights when motion is detected in a certain area.
- **Home automation**: Turn on lights or start a device when a person enters a room.
- **Robotics**: Enable robots to navigate their environment by reacting to motion, helping them avoid obstacles.

Example project: Create a **motion-activated security system** that turns on a light or triggers an alarm whenever motion is detected. By integrating a PIR sensor with Arduino, the system can detect motion and respond instantly, providing real-time security for your home or office.

How Sensors Make Your Projects More Dynamic

When you combine these sensors into your projects, you create a more **interactive and responsive system**. The key to making a project truly dynamic is **feedback**—having the system react based on the sensory data it receives.

Here are a few examples of how sensors can make your projects more dynamic:

1. Creating Smart Home Devices

By using **motion**, **light**, and **temperature** sensors, you can build a fully automated home system that reacts to your environment. For example, lights could turn on when motion is detected, and fans could activate when the temperature exceeds a certain level. This allows your home to automatically adjust itself to your needs, improving convenience and energy efficiency.

2. Building Interactive Robots

Sensors can help robots interact with their environment in real-time. For instance, a **proximity sensor** can help a robot avoid obstacles, while a **light sensor** can enable it to move toward light sources. By adding **motion** and **temperature sensors**, you can build robots that adapt to their surroundings, avoid hazards, and respond to changes in the environment.

3. Enhancing Learning and Exploration

Sensors enable Arduino projects to become more hands-on and engaging. For example, in a science fair project, students could create a model of a greenhouse that automatically adjusts humidity and temperature based on sensor readings, or a weather station that logs and tracks temperature and light changes throughout the day. These projects bring abstract concepts to life and make learning more immersive and interactive.

Practical Example: Dynamic Lighting System

Let's illustrate how sensors can create dynamic, responsive projects by building a simple **dynamic lighting system**. In this system, a light sensor and a temperature sensor work together to control the lighting in a room:

- The **light sensor** detects the ambient light level, and if it's below a certain threshold (e.g., during the evening or cloudy weather), it will trigger the Arduino to turn on the lights.

- The **temperature sensor** detects the room's temperature, and if it gets too hot or too cold, it can turn the lights off or trigger a fan.

By combining **two sensors** with **two actuators** (the light and fan), you create an interactive system that dynamically responds to changes in the environment, automating the process of controlling lighting and temperature.

Practical Examples: Going Beyond Theory with Simple Projects Using Sensors

As you embark on your journey into Arduino programming, it's essential to move beyond the theoretical concepts and start building real, tangible projects. Working with **sensors** opens up a world of possibilities where you can create interactive systems that respond to the physical world in real-time. To help bring these concepts to life, we'll walk through a **simple, hands-on project**—a **motion-sensing light system**—using basic sensors. This example will show you how easy it is to use **motion sensors** to automate everyday tasks, such as turning on a light when someone enters a room.

Project Overview: Motion-Sensing Light System

In this project, we'll use a **PIR (Passive Infrared) motion sensor** to detect movement in a room and trigger an **LED light** to turn on when motion is detected. When no motion is sensed for a set period, the light will turn off. This is a great starting point for understanding how sensors can make your projects more dynamic and responsive to real-world changes.

What You'll Need for This Project:

- **Arduino board** (e.g., Arduino Uno)
- **PIR motion sensor** (commonly used for detecting motion based on infrared radiation)
- **LED light** (for indicating whether motion has been detected)
- **Resistor** (220 ohms for the LED)
- **Breadboard and jumper wires**
- **Arduino IDE** (for writing and uploading code)

How the Motion-Sensing Light Works:

1. The **PIR motion sensor** detects infrared radiation from a warm object, such as a human body. When someone enters the sensor's detection range, it sends a **HIGH signal** to the Arduino, which triggers the light to turn on.
2. If the sensor detects no motion for a certain period, the Arduino sends a **LOW signal** to the LED, turning the light off.

Step-by-Step Instructions:

Step 1: Wiring the Components

1. **Connect the PIR sensor**:
    - Connect the **VCC pin** of the PIR sensor to the **5V pin** on the Arduino.
    - Connect the **GND pin** of the PIR sensor to the **GND pin** on the Arduino.
    - Connect the **OUT pin** of the PIR sensor to **digital pin 7** on the Arduino.
2. **Connect the LED**:

- Connect the **longer leg** (anode) of the LED to **digital pin 13** on the Arduino.
- Connect the **shorter leg** (cathode) to one side of the **220-ohm resistor**.
- Connect the other side of the resistor to **GND** on the Arduino.

Hands-On Activity: Experimenting with Different Sensors

Now that you've seen how to use a **motion sensor** to create an interactive project, let's explore how you can extend this knowledge by experimenting with other types of sensors and integrating them into your own personal projects.

Activity 1: Light-Sensing LED Control

Instead of using a motion sensor, let's explore how a **light sensor** (like a **photoresistor** or **LDR**) can be used to control an LED. This will allow you to create a simple **automatic light system** that turns on or off based on the ambient light level.

Materials Needed:

- **LDR (Light Dependent Resistor)**
- **10k Ohm resistor** (for the voltage divider circuit)
- **LED and 220-ohm resistor**
- **Arduino board**

Steps:

1. **Set up the LDR**: Connect the LDR to **5V** and **GND**, with a **10k Ohm resistor** in between to create a voltage divider circuit.

2. **Connect the LED**: Connect the LED to **pin 13** as you did before.
3. **Code**: Modify the code to read the light level from the LDR and control the LED based on the ambient light.

This project is an excellent way to understand how light sensors work and how they can be used to make your projects respond to the environment in a meaningful way.

Activity 2: Temperature-Activated Fan

Another practical project idea is to create a **temperature-controlled fan** using a **DHT11 temperature sensor**. This system can measure the temperature in a room and automatically turn on a fan if the temperature exceeds a certain threshold.

Materials Needed:

- **DHT11 temperature sensor**
- **Fan** (controlled via a transistor or relay)
- **Arduino board**

Steps:

1. **Set up the DHT11 sensor**: Connect the sensor to the Arduino and read temperature data.
2. **Write Code**: Program the Arduino to turn on the fan when the temperature exceeds a preset value (e.g., 25°C).
3. **Test the System**: See how the fan activates and deactivates based on the temperature in the room.

This activity helps you understand how to use temperature sensors and actuators like fans to control physical devices based on environmental data.

Encouraging Exploration and Creativity

As you continue experimenting with sensors, I encourage you to think about how they can be integrated into your **own personal projects**. Here are a few ideas to get you started:

- **Smart plant watering system**: Use a **moisture sensor** to detect when your plant needs water and trigger a pump to water it automatically.
- **Security system**: Combine **motion sensors** and **sound sensors** to detect movement and sound, triggering an alarm or camera system.
- **Interactive art installations**: Use sensors to create art that changes or reacts to the audience's movements or environmental conditions.

The possibilities are endless when you combine sensors with the creativity of Arduino programming. Experiment with different types of sensors, mix and match components, and see how you can automate everyday tasks or create fun, interactive systems. Each project you build will expand your understanding of sensors, giving you the confidence to tackle even more complex and innovative ideas.

Chapter 5: Coding Fundamentals: How Arduino Makes Coding Easy for Everyone

Coding can seem like a daunting task, especially for those who have never written a line of code before. The world of programming is filled with complex syntax, intricate logic, and confusing jargon. However, **Arduino** makes it easy for beginners to dive into the world of coding without the overwhelming complexity that usually accompanies it. Whether you're aiming to create a simple project or something more advanced, Arduino offers an accessible entry point that empowers everyone—regardless of age or experience level—to turn their ideas into reality.

In this chapter, we'll break down the core **coding fundamentals** behind Arduino programming. By the end of this chapter, you'll not only understand how Arduino works but also feel confident writing basic programs (known as **sketches**) and experimenting with them on your own. We'll simplify complex concepts and show you how Arduino's unique features make it beginner-friendly.

What is Arduino Programming?

Arduino programming is based on **C++**, a powerful, flexible, and widely-used programming language. However, unlike traditional C++ programming, Arduino abstracts much of the complexity away, making it more accessible for beginners. **Arduino sketches** (the name for programs written for Arduino) are a simplified version of C++ that allow you to easily control hardware like sensors, LEDs, motors, and more.

Arduino's ease of use stems from its **Integrated Development Environment (IDE)**, a user-friendly platform that helps you write, compile, and upload your sketches to the Arduino board. The IDE is designed to be straightforward and intuitive, allowing you to focus on learning how to code, not getting bogged down by the intricacies of programming languages.

Arduino Programming Structure: The Building Blocks of Code

Let's take a closer look at how an Arduino sketch is structured. Every Arduino sketch contains at least two functions that make up the skeleton of the program:

1. Setup Function: void setup()

The setup() function is the first piece of code that runs when the Arduino is powered on or reset. This function is typically used to **initialize variables**, **configure hardware pins**, and set up other necessary conditions that will be used throughout the program.

In this function, you set the **input** and **output** pins, specify communication protocols (e.g., serial communication), and perform any other setup tasks that need to be done once at the start.

2. Loop Function: *void loop()*

After the setup() function runs, the **loop()** function takes over. It continuously runs in a cycle, allowing your Arduino to **constantly check conditions** and **perform actions**. The code inside the loop() function runs repeatedly until the Arduino is powered off or reset.

This is where most of your program logic will go, such as **reading sensors**, **controlling actuators**, or **checking user inputs**. The loop function ensures that your Arduino is always actively working.

In this example, the LED connected to pin 13 will blink on and off every second.

Basic Arduino Syntax: Understanding the Language

To start coding with Arduino, you don't need to memorize every C++ command out there. The Arduino IDE simplifies things, and the structure of an Arduino program is built on only a few key elements:

1. Variables and Data Types

In any programming language, **variables** are used to store data that can change over time. When coding in Arduino, you need to declare variables that represent things like sensor readings, pin numbers, or states of devices (e.g., whether an LED is on or off).

Here are some common **data types** you'll use with Arduino:

- int: Used to store whole numbers (e.g., 1, 25, 100).
- float: Used to store decimal numbers (e.g., 3.14, 0.99).
- bool: Used to store true/false values (e.g., HIGH or LOW, true or false).
- char: Used to store a single character (e.g., 'A', 'Z').

2. Functions

A **function** is a block of code that performs a specific task. Functions help you break up a program into manageable chunks, which makes your code more readable and reusable.

In Arduino, you can use built-in functions, such as digitalWrite() and delay(), or create your own functions to simplify your program.

Example of a custom function:

```cpp
void blinkLED() {
  digitalWrite(ledPin, HIGH);
  delay(1000);
  digitalWrite(ledPin, LOW);
  delay(1000);
}
```

You can call this function in your loop() function whenever you want to blink the LED.

3. Control Structures

Control structures like **if statements**, **loops**, and **switch statements** are essential to deciding what your program should do under different conditions. These are the "brains" of your code, allowing it to react to inputs or make decisions.

- if: Executes code if a condition is true.
- else: Executes code if the condition is false.
- for/while: Loops that repeat code multiple times.

Making Coding Accessible: How Arduino Simplifies Programming for Everyone

What makes Arduino so beginner-friendly is its unique combination of **simplified syntax**, **visual feedback**, and **hands-on learning**. Here's how Arduino makes coding easier for everyone:

1. Immediate Results with Real-World Feedback

When you write a program for Arduino, the results are immediate. Unlike traditional software programming, where you might not see the results until the program is complete, with Arduino, you can upload your code and see the results **instantly**. Want to control an LED? Write the code and see the light blink! Want to read a temperature sensor? See the data in the Arduino IDE.

This immediate feedback builds confidence and reinforces the learning process, making it much easier to stay motivated and understand how your code is working.

2. Extensive Documentation and Tutorials

Arduino has a vast, supportive community, and an extensive collection of **tutorials**, **example codes**, and **documentation** to help beginners. No matter the project you're working on, you'll find plenty of resources to guide you through the process.

3. Open-Source Community and Libraries

Arduino's open-source nature means that thousands of libraries are available to extend the functionality of your projects. These libraries simplify complex tasks like controlling motors, reading sensors, or connecting to Wi-Fi. Rather than writing complex code from scratch, you can simply include a library and focus on the high-level design of your project.

For example, the **Servo library** makes it easy to control servos without having to understand the low-level code needed to generate the correct PWM (Pulse Width Modulation) signals.

4. Hardware + Software Integration

Unlike traditional coding, where you only interact with virtual objects, Arduino allows you to connect **hardware** and **software**. This integration creates a more tangible learning experience, as you're working with real sensors, actuators, and devices. This hands-on approach accelerates learning and reinforces abstract coding concepts.

Simplifying Coding Concepts: Essential Building Blocks for Arduino Programming

Coding is often seen as a complex skill, filled with abstract concepts and mysterious syntax. But when broken down into bite-sized chunks and explained through relatable analogies, it becomes easier to understand and more intuitive to use. In this section, we'll demystify essential coding concepts like **loops**, **variables**, and **functions**, and connect them to real-world scenarios, making them easier to grasp. Visual aids and everyday examples will help you see how these fundamental building blocks of Arduino programming are similar to things you already know and use.

1. Loops: Repeating Actions Over Time

What is a Loop?

At its core, a **loop** is a way to make something happen repeatedly—over and over again—until a certain condition is met. In programming, loops are used to automate repetitive tasks without having to write the same instructions multiple times. Think of loops as the **repetitive steps** you might follow in your daily life.

Real-World Analogy:

Imagine you're preparing for a party, and you need to set up decorations. You decide to put up a string of lights across the room. Instead of manually going around and placing each light individually in different spots, you simply follow a pattern: **take a light, hang it, and repeat** until the entire room is decorated.

In programming, this pattern is similar to how a **loop** works: you define the action (hanging a light), and the loop keeps doing it until a specified condition is met (when the room is fully decorated).

Arduino Example:

In Arduino, loops are written inside the loop() function, and they execute repeatedly as long as the Arduino is powered on. Here's an example where an LED is turned on and off in a loop.

This loop runs endlessly, turning the LED on for 1 second, then off for 1 second, over and over again.

2. Variables: Storing Information for Later Use
What is a Variable?

A **variable** in programming is like a container or a box where you can store data that may change over time. Variables hold information, such as numbers, text, or other values, and can be referenced or altered later in the program. In Arduino programming, variables are essential for making your project interactive and responsive.

Real-World Analogy:

Think of a **variable** as a **post-it note**. You can write something down on it, and at any point, you can look at it, change what's written, or move it around. For example, you may write down your **current temperature** on a note and refer to it as you monitor how hot or cold a room is. Later, you can update the temperature on the note without having to rewrite the entire program.

In the same way, variables in programming hold values, like a number or a string, that you can change or check whenever needed.

Arduino Example:

When you work with Arduino, you might want to store data like a sensor reading or a pin state in a variable. For example, to store the status of a sensor, you might use a variable:

Here, the `sensorValue` variable holds the reading from the sensor and is updated each time the loop runs. Just like you might write a new number on your post-it note, the sensorValue variable gets updated with the new data.

3. Functions: Organizing Code into Manageable Tasks

What is a Function?

A **function** in programming is a way to group a set of instructions into a single, reusable block of code. It allows you to **organize your program** and **simplify repetitive tasks** by defining the code once and calling it whenever needed. Think of a function as a **tool or machine** that performs a specific job.

Real-World Analogy:

Imagine you run a bakery, and you need to bake cupcakes every day. Rather than going through the entire process of mixing, pouring, and baking each time from scratch, you set up a machine (or a recipe) that does it for you. Whenever you want to bake cupcakes, you simply press a button to start the machine, which will go through the same set of steps every time.

In the same way, a **function** is like that machine or recipe: once you've created it, you can "call" it whenever you need it, and it will repeat the steps inside without you needing to rewrite the code each time.

Visual Aids: Making Coding Concepts Clearer

1. Flowcharts for Loops:

A flowchart can be a helpful visual aid for understanding how a loop works. Here's a simple flowchart for the **blinkLED()** loop:

sql

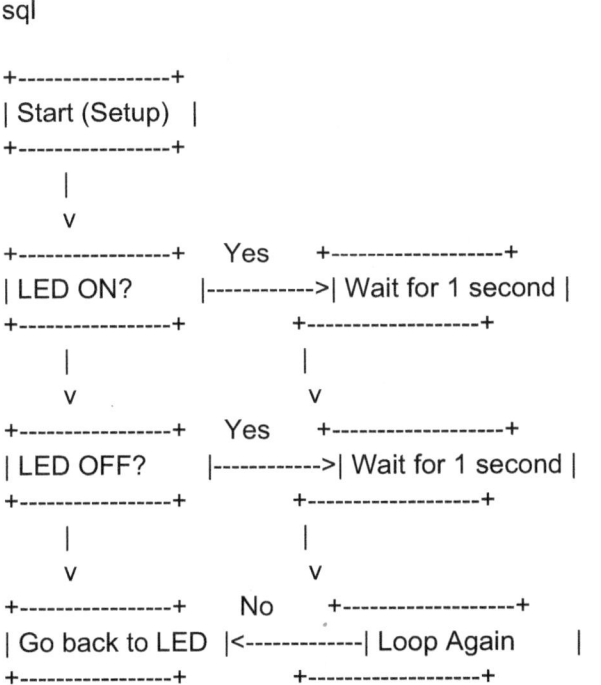

This shows that the loop continually checks whether to turn the LED on or off, and once it finishes one cycle (LED on, wait, LED off, wait), it starts again.

2. Diagrams for Functions:

A visual representation of a **function** can help you see the flow of how the function works:

scss

```
+------------------------+
| Start of blinkLED()    |
+------------------------+
           |
           v
+------------------------+   (turns LED on)
| digitalWrite(13, HIGH) |
+------------------------+
           |
           v
+------------------------+   (wait 1 second)
| delay(1000)            |
+------------------------+
           |
           v
+------------------------+   (turns LED off)
| digitalWrite(13, LOW)  |
+------------------------+
           |
           v
+------------------------+   (wait 1 second)
| delay(1000)            |
+------------------------+
           |
           v
+------------------------+
| End of blinkLED()      |
+------------------------+
           |
           v
+------------------------+
| Back to main program   |
+------------------------+
```

Breaking Down the Barrier to Learning Code

By simplifying coding concepts such as **loops**, **variables**, and **functions** through relatable analogies and visual aids, we can make the process of learning Arduino programming more accessible and enjoyable. These fundamental building blocks form the foundation for all your Arduino projects, and understanding them deeply will help you become more confident and effective in your coding journey.

Interactive Challenges: Reinforcing Learning Through Hands-On Experimentation

One of the most effective ways to learn Arduino programming (or any programming language) is through **active experimentation**. It's not enough to simply read through code and understand what it does — the real learning happens when you roll up your sleeves and modify the code yourself. By working through **interactive challenges**, you can reinforce the concepts and gain a deeper understanding of how everything fits together.

In this chapter, we'll guide you through a series of **interactive exercises** that will help you modify existing code, experiment with different inputs and outputs, and discover new ways to achieve similar results. These challenges will encourage you to apply your knowledge, troubleshoot errors, and ultimately become more confident and creative in your approach to Arduino programming.

1. Challenge 1: Adjusting LED Blink Timing
Objective:

This challenge focuses on modifying the **delay** times in the LED blinking project. By experimenting with different time intervals, you'll see how the code affects the LED's behavior and get hands-on experience adjusting a key parameter in Arduino programming.

Instructions:

You'll start with the basic **Blinking LED** code, which turns an LED on and off at regular intervals. Your task is to modify the code to change the timing, so the LED blinks faster or slower.

Steps to Experiment:

1. **Increase the blink speed:**
   - Modify the delay() function from 1000ms to 500ms to make the LED blink twice as fast.
   - **New Code:**

   cpp

   delay(500); // Wait for 0.5 seconds

2. **Slow down the blink:**
   - Try changing the delay() to 2000ms for a slower blink, making the LED blink every two seconds.
   - **New Code:**

   cpp

   delay(2000); // Wait for 2 seconds

3. **Experiment with different patterns:**
   - You could modify the code to make the LED blink faster for a while, then slower after a certain time.
   - For example, blink fast for the first 10 seconds, then slow down:

Learning Outcome:

By experimenting with the delay times, you'll better understand how **timing** affects the behavior of electronic components like LEDs. This exercise also reinforces the importance of the delay() function in controlling the flow of code execution, as well as how to modify parameters to adjust output.

2. Challenge 2: Modifying the LED Blink Pattern
Objective:

This challenge builds on the previous one, but instead of adjusting the timing, you will focus on creating different **blink patterns** using multiple LEDs. You'll modify the code to control more than one LED, and experiment with different patterns like alternating or simultaneous blinking.

Instructions:

To do this challenge, you'll need to connect multiple LEDs to different digital pins (e.g., pins 13, 12, and 11). Your task is to create a blink pattern that alternates between the LEDs.

1. **Create a custom pattern:**
   - Write a sequence that makes LEDs blink in a custom pattern, such as a "chase" pattern or a "ping-pong" effect where the LEDs blink in sequence from left to right or right to left.

Learning Outcome:

This exercise helps you understand how to manage multiple outputs in a circuit and control the behavior of each individually. It also reinforces the concept of **logical flow** and **timing** in code. You'll see how different sequences can create visually engaging results and how controlling multiple pins can create more complex interactions.

3. Challenge 3: Using a Button to Control an LED
Objective:

This challenge allows you to combine **input** and **output** by controlling the LED with a **button**. The task is to write code that reads the button's state and turns the LED on when the button is pressed, and off when it's released.

Instructions:

You'll need to connect a push button to a digital input pin (e.g., pin 2), and an LED to a digital output pin (e.g., pin 13).

Steps to Experiment:

1. **Change the LED behavior:**
   - Modify the code so the LED blinks instead of turning on when the button is pressed.
   - **Example:** Use delay() and digitalWrite() inside the if statement to blink the LED when the button is pressed.
2. **Add multiple buttons:**
   - Use two or more buttons to control different LEDs. For example, one button could turn on an LED, while another turns off a different LED.
3. **Debounce the button:**
   - Modify the code to debounce the button press so that multiple presses don't register as one.

This will prevent unexpected behavior from the button's electrical noise.

Learning Outcome:

This challenge gives you the hands-on experience of **reading input** from a sensor (the button) and using that input to control an output (the LED). By experimenting with different inputs and outputs, you'll develop a deeper understanding of **input/output control**, **logical decisions**, and the importance of user interaction in Arduino projects.

A Focus on Repetition: Providing Multiple Ways to Solve Problems Using Different Coding Structures

Repetition is one of the most powerful learning tools in programming. It's through repeated exposure and practice that complex concepts become second nature. When you're learning Arduino programming, repetition goes beyond simply executing the same task multiple times. It's about solving problems in different ways, using various coding structures, and gaining a deeper understanding of how each approach works.

In this section, we'll explore how you can tackle problems from multiple angles by applying different coding structures — such as **loops**, **conditional statements**, and **functions** — to solve the same problem. Each structure offers a unique way of thinking about and solving the problem, helping you build flexibility and confidence as you progress. By experimenting with these structures, you will gain a more profound understanding of how and when to use each of them.

1. Repetition with For Loops

A **for loop** is one of the most common structures in programming, ideal for situations where you need to repeat a task a specific number of times. Let's look at how you can use a for loop to solve a problem in Arduino programming.

Problem: Blink an LED five times

The objective is simple: make an LED blink five times with a one-second delay between each blink. First, let's see how you would solve this using a for loop.

Explanation:

- The **for loop** in this code runs 5 times, turning the LED on and off with a delay between each action.
- The loop starts with i = 0 and increments i until i equals 5, meaning it will run five times.

Why this is beneficial:

Using a for loop here makes it clear that repetition is occurring a specific number of times. It simplifies the code, reducing redundancy and enhancing clarity.

2. Repetition with While Loops

Another way to achieve repetition is with a **while loop**, which repeats a block of code as long as a specified condition is true. Unlike a for loop, the number of iterations is not fixed ahead of time but depends on the condition being met.

Problem: Blink an LED until a button is pressed

This time, instead of repeating a set number of times, the goal is to continuously blink an LED until a button is pressed. Here's how we can solve it using a while loop.

Explanation:

- In this example, the while loop continues to execute as long as the button is not pressed (buttonState == LOW).
- The loop checks the button's state on every iteration. When the button is pressed, the loop exits, and the LED stops blinking.

Why this is beneficial:

The while loop is particularly useful for scenarios where you don't know in advance how many times an action needs to be repeated — you only know that the repetition should continue while a certain condition holds true.

3. Repetition with Functions

Another powerful way to achieve repetition is by using **functions**. Functions allow you to group code into reusable blocks. This means you can call the same function multiple times throughout your code without having to rewrite the same logic repeatedly.

Problem: Blink an LED multiple times at different intervals

This time, instead of writing the blinking code directly in the loop(), you can encapsulate it inside a function and call that function whenever needed.

Explanation:

- In this code, the function blinkLED() is called twice within loop(). Each time it's called, it blinks the LED a different number of times, with different delay intervals.
- By defining the behavior inside a function, you avoid repeating the same block of code and can reuse it wherever needed.

Why this is beneficial:

Functions let you write more efficient, modular code. You define a task once and can call it many times, which makes your code cleaner, easier to read, and less error-prone. Functions also encourage **code reuse**, helping you solve problems without duplicating effort.

4. Repetition with Arrays and Loops

Now, let's dive into **arrays**, which are a great way to manage and manipulate a collection of related data. Arrays can hold multiple values (e.g., multiple LEDs or sensors), and you can use a loop to repeat an action on each element in the array.

Problem: Blink multiple LEDs in sequence

You can use an array to store the pin numbers of multiple LEDs, and then use a for loop to control each LED in sequence.

Explanation:

- Here, the ledPins[] array stores the pin numbers of the LEDs. The for loop iterates over the array and controls each LED in sequence.

- This method avoids repeating the same code for each LED pin individually and demonstrates how repetition can be efficiently managed using arrays.

Why this is beneficial:

Arrays allow you to group related data together, and using loops with arrays makes your code more scalable. You can easily add or remove LEDs from the array, and the loop will automatically adjust.

Chapter 6: Introduction to Robotics: The Fascinating World of Arduino Robots

The world of robotics is a fascinating frontier where science fiction meets reality, and **Arduino** plays a pivotal role in making this cutting-edge field accessible to everyone — from hobbyists and students to professional engineers and inventors. Arduino's open-source hardware and software provide an affordable and user-friendly platform to build robots that can perform a variety of tasks, from simple movements to complex decision-making processes. In this chapter, we will dive into the fundamentals of robotics and explore how **Arduino** is revolutionizing the way we design, build, and control robots.

The Rise of Robotics in Everyday Life

Robotics is no longer a distant concept or something confined to factories or laboratories. From self-driving cars to robotic vacuum cleaners and even humanoid robots, robotics has become a part of our everyday lives. As technology advances, robots are being integrated into industries ranging from healthcare to manufacturing, entertainment, and even home automation. The exciting part is that you don't need to be an expert engineer or spend thousands of dollars to start building your own robots — **Arduino** is here to make robotics accessible to anyone with an interest in technology and problem-solving.

What Makes Arduino Ideal for Robotics?

Arduino is a versatile and inexpensive platform that empowers users to bring their ideas to life. So, why is Arduino such an ideal choice for building robots?

1. **Affordability:**
   Arduino boards and components are relatively inexpensive, making it easy for beginners to get started with robotics without breaking the bank. This lowers the barrier to entry and opens the door to experimentation and innovation.
2. **Open Source:**
   Arduino is open-source, which means the design files, software, and schematics are available for anyone to use, modify, and improve. This creates a vast community of makers, engineers, and hobbyists who continuously contribute to the development of new ideas and resources.
3. **Wide Range of Components:**
   Arduino is compatible with a variety of sensors, actuators, motors, and communication devices that are essential for building robots. This flexibility makes Arduino an excellent platform for creating robots with varying degrees of complexity.
4. **Simplicity and Flexibility:**
   The Arduino IDE (Integrated Development Environment) is straightforward to use, even for beginners. You don't need advanced knowledge of programming or hardware design to start building functional robots. Arduino's flexibility also means you can easily scale your projects from simple designs to more sophisticated robotic systems.

Essential Robotics Concepts

Before we dive into building our first robot, it's important to understand some basic concepts that will serve as the foundation of your robot-building journey.

1. Actuators: The Movers of Your Robot

Actuators are the parts of your robot that allow it to move or interact with its environment. They convert electrical signals into physical movement. In Arduino-based robotics, the most common types of actuators are:

- **Motors:** These include DC motors, servo motors, and stepper motors, which allow robots to move in various ways (e.g., driving, rotating, or precisely positioning parts).
- **Servos:** These are specialized motors that can rotate to specific angles, often used in robotic arms or grippers.

2. Sensors: Helping Robots Perceive Their Environment

Sensors give your robot the ability to understand its surroundings and respond accordingly. They act as the robot's "eyes," "ears," and "nervous system." Some of the most common sensors you can use in Arduino-based robots include:

- **Ultrasonic Sensors:** Used for distance measurement, enabling robots to detect obstacles and avoid collisions.
- **Infrared Sensors:** Often used for line-following robots, IR sensors can help robots track a path or avoid objects by sensing light reflection.

- **Touch Sensors:** These sensors can detect physical contact and are often used to trigger actions, such as stopping or changing direction.
- **Gyroscopes and Accelerometers:** These sensors are used to measure orientation and motion, helping robots balance or determine their position in space.

3. Power Supply: Giving Life to Your Robot

A reliable power supply is essential for your robot to function. Depending on your robot's needs, you may use:

- **Batteries (AA, Li-Po, etc.):** The most common power source for Arduino robots, providing portable energy.
- **External Power Sources:** For larger robots that require more power, external supplies may be used to ensure consistent and stable performance.

Arduino in Robotics: A Step-by-Step Guide to Building Your First Robot

Now that we have covered the basic concepts, let's explore how you can start building your first Arduino-powered robot. We will use an example of a **simple line-following robot** — a robot that uses sensors to detect a line and follow it autonomously.

Basic Components for Your First Arduino Robot:

- **Arduino Uno or any compatible board**
- **Two DC motors** and a motor driver (such as L298N)
- **Ultrasonic sensor** for distance sensing (optional for more advanced projects)
- **Infrared sensors** for line detection

- **Chassis** for your robot (can be a pre-made one or custom-built)
- **Wheels and tires** for movement
- **Battery pack** for power
- **Jumper wires and connectors**

Step 1: Assemble the Robot's Structure

1. **Build the Chassis:**
   If you're using a pre-made chassis, attach the DC motors to the chassis and mount the wheels. If you're using a custom-built chassis, ensure it has enough space to house your Arduino board, sensors, and battery pack.
2. **Mount the Motors:**
   Attach the motors to the chassis and connect them to the motor driver module (such as L298N). The motor driver allows the Arduino to control the direction and speed of the motors.
3. **Install the Sensors:**
   Place the infrared sensors on the bottom of the robot to detect the line. Position them so they can detect the contrast between the line (usually black) and the surface (usually white).
4. **Connect the Arduino:**
   Secure the Arduino board on the chassis and wire it to the motor driver, sensors, and power supply.

Step 3: Test and Improve Your Robot

Once the hardware and code are set up, it's time to test your robot. Place it on a track with a clearly marked line, and watch as it follows the path.

Troubleshooting Tips:

- **Sensor calibration:** Make sure your sensors are correctly positioned and calibrated to detect the line.
- **Motor connections:** Double-check that the motor connections to the motor driver are correct.
- **Power issues:** Ensure your power supply is sufficient to run the motors and Arduino.

Expanding Your Knowledge: Beyond the Line-Follower

After building a simple line-following robot, the possibilities for Arduino-based robotics are endless. You can build robots that:

- **Avoid obstacles** using ultrasonic sensors.
- **Follow voice commands** with Bluetooth and a smartphone app.
- **Perform specific tasks** like picking up objects or interacting with their environment.
- **Learn from their surroundings** with machine learning and artificial intelligence.

Arduino opens up a world of possibilities in robotics, giving you the tools to create your own inventions, solve real-world problems, and bring your creative ideas to life. Whether you're interested in making robots for fun, as a learning project, or as part of a professional career in robotics, Arduino provides an accessible platform for all.

One of the most powerful aspects of Arduino is its ability to turn abstract ideas into tangible reality. With Arduino, the possibilities for building robots are endless — and the best part is, you don't need to be a professional engineer to start creating. Whether you're just starting or looking to expand your robotics knowledge, Arduino allows anyone to build functional robots with relative ease. Let's explore a few simple projects you can create with Arduino to inspire your imagination and spark your innovation.

1. Basic Line-Following Robot

A classic beginner robotics project is the **line-following robot**. This robot uses infrared (IR) sensors to detect a line (usually black) on a light surface (typically white) and follow it autonomously. This project is an excellent way to understand how sensors interact with the environment and how robots can process data to make decisions.

How It Works: The robot uses two sensors placed on the ground to detect the line. If the sensors detect that both sides are on the line, the robot moves forward. If one sensor detects the line while the other doesn't, the robot adjusts its direction to stay on course.

Components Needed:

- Arduino board (Uno, Nano, etc.)
- IR sensors (left and right)
- DC motors and wheels
- Motor driver (L298N or similar)
- Chassis (or a basic frame)
- Battery pack

Project Benefits:

- Teaches you about sensor integration and motor control.

- Provides hands-on experience with decision-making logic in programming.
- Demonstrates how simple robots can be made to interact with their environment.

This project is an ideal starting point for anyone new to robotics because it shows how sensory data can be interpreted by a microcontroller (Arduino) to control physical actions (movement of motors). By successfully completing this project, you will gain foundational skills that can be expanded into more complex robotic systems.

2. Obstacle-Avoidance Robot

Building an **obstacle-avoidance robot** is a fantastic next step after your line-following project. This robot uses an ultrasonic sensor to detect objects in its path and automatically adjusts its direction to avoid them. This introduces the concept of navigation and environment interaction, expanding your understanding of robotics.

How It Works: The ultrasonic sensor emits sound waves and measures the time it takes for the sound to bounce back from an obstacle. If an obstacle is detected within a specific range, the robot changes its direction to avoid the obstacle.

Components Needed:

- Arduino board
- Ultrasonic sensor (HC-SR04)
- DC motors with wheels
- Motor driver
- Chassis
- Battery pack

Project Benefits:

- Teaches you about distance measurement and sensor integration.
- Expands your understanding of robot autonomy and decision-making processes.
- Offers insight into how robots can adapt to unexpected changes in their environment.

3. Robotic Arm

A **robotic arm** is a more advanced project that introduces the concept of precise movements and actuator control. By using servo motors, this robot arm can mimic basic human hand movements such as picking up and placing objects.

How It Works: A robotic arm typically consists of several joints (servo motors), each allowing for movement in a specific direction. By sending signals to the servos, you can control the arm to perform different tasks, like grabbing or moving objects.

Components Needed:

- Arduino board
- Servo motors (3 or 4 for a simple arm)
- Gripper mechanism (could be a simple 2-finger gripper)
- Chassis/frame to mount the servos
- Power supply

Project Benefits:

- Introduces advanced concepts in robotic kinematics.

- Provides practical experience with servo motor control and multi-axis movement.
- Teaches how a robot can execute tasks with precision.

4. Smart Robot Car

A **smart robot car** integrates multiple sensors and actuators to create a robot car capable of making decisions based on its surroundings. By using a combination of ultrasonic sensors, cameras, and even GPS modules, this project can make a robot car navigate a room or environment autonomously.

How It Works: The robot car uses sensors to detect obstacles, track a path, or even map out an area. Depending on the complexity of the project, you can add features like remote control via Bluetooth or Wi-Fi, allowing you to control the robot car from your smartphone.

Components Needed:

- Arduino board
- Ultrasonic sensors, IR sensors, or cameras for navigation
- DC motors and motor driver
- Chassis and wheels
- Battery pack
- Bluetooth/Wi-Fi module (optional for remote control)

Project Benefits:

- Teaches you about robotics, navigation, and autonomy.
- Expands your skills in wireless communication and remote control.

- Provides insight into how autonomous vehicles and robots are designed for real-world use.

Focus on Practical Use: How Robotics Solves Real-World Problems

While building robots is exciting, it's important to understand how robotics can address practical issues and be applied to real-world challenges. Robotics is more than just a fun hobby or academic exercise — it has real-world applications that can improve industries, everyday life, and personal hobbies. Let's explore how robotics can solve some of the most pressing problems in the world today.

1. Automation in Industry

In manufacturing, robotics has revolutionized the way goods are produced. Robots can automate repetitive tasks, handle dangerous materials, and perform precise movements at high speeds.

- **Assembly lines:** Robots are now common on factory floors, where they can assemble components faster and more accurately than humans.
- **Inspection and quality control:** Robots can also perform quality checks, ensuring products meet high standards.
- **Material handling:** Autonomous robots can move materials from one part of the factory to another, improving efficiency and safety.

By learning to build robots with Arduino, you can better understand the inner workings of industrial robots and how they help improve productivity and safety in factories worldwide.

2. Healthcare and Medicine

Robots are also making a huge impact in the healthcare sector. **Surgical robots**, for example, enable doctors to perform minimally invasive surgeries with greater precision and control. They allow for smaller incisions, less recovery time, and fewer complications. Additionally, robots are being developed to assist with elder care, from robotic assistants helping seniors with daily tasks to robotic exoskeletons that assist in rehabilitation.

- **Telemedicine:** Robots enable remote surgeries or consultations, making healthcare more accessible to people in remote areas.
- **Robot-assisted surgery:** Precision instruments guided by robots can be used for high-risk surgeries, allowing surgeons to make smaller incisions and avoid damaging healthy tissues.

By understanding how to program and build robots, Arduino enthusiasts can contribute to the growing field of medical robotics, improving patient outcomes and increasing accessibility to healthcare.

3. Personal Hobbies: Enhancing Creativity and Fun

Robotics isn't only for industrial applications — it's also a fantastic way to express creativity and have fun. You can design and build robots that do everything from playing games to creating art.

- **Home automation:** Smart home robots can automate tasks like adjusting lights, controlling the thermostat, or managing security systems. By incorporating sensors and smart technology, you can create robots that make your life easier and more comfortable.
- **Entertainment robots:** Some robots can interact with their surroundings in creative ways, like dancing to music, playing games, or even creating art.
- **DIY projects:** Building a simple robot for personal use can be incredibly rewarding. Whether it's a robotic assistant that helps with small tasks or a simple robot that follows a light or sound, these projects allow you to combine engineering skills with personal interests.

Robotics is a versatile field that touches almost every aspect of daily life, from entertainment to education to practical applications. By understanding the basics of Arduino robotics, you can start solving real-world problems, automating your life, or just having fun with technology.

Collaborative Learning: Joining the Global Arduino Community

One of the most exciting aspects of learning Arduino and robotics is the opportunity for **collaborative learning**. While the process of building projects on your own can be fulfilling, being part of a community of like-minded hobbyists and experts opens doors to a wealth of knowledge, inspiration, and problem-solving. Whether you're looking to improve your skills, solve technical challenges, or share your latest project, engaging with others can accelerate your learning and keep you motivated.

In this section, we'll explore the many online communities and resources that can help you grow as an Arduino enthusiast. These communities are perfect for getting feedback on your projects, learning from others, and even collaborating on large-scale initiatives.

Why Collaborative Learning is Powerful

Collaborative learning involves engaging with others to share knowledge, troubleshoot problems, and explore new ideas. When you join an Arduino community, you'll have access to a diverse range of perspectives. Beginners and experts alike contribute, and the shared learning environment helps everyone improve.

Here are a few reasons why **collaborative learning** can make you a better Arduino builder:

1. **Diverse Problem-Solving**: You may come across challenges that seem impossible to solve alone. Community members with different skill sets may offer solutions or suggest workarounds that you hadn't considered.
2. **Motivation and Inspiration**: Seeing others' projects can inspire new ideas. A community offers opportunities to learn from exciting and innovative creations, which helps keep your enthusiasm alive.
3. **Instant Help and Feedback**: Instead of spending hours troubleshooting, you can ask the community for help. You're more likely to get quicker, real-time assistance, saving time and avoiding frustration.
4. **Collaborative Projects**: Working on projects together can enhance your skills and introduce you to new techniques. Collaborative projects also expose you to

areas of Arduino that you might not have explored on your own.

Now, let's take a look at some of the best online resources where you can engage with the Arduino community.

1. Arduino Official Website & Forum

The **Arduino website** itself is a great starting point for anyone interested in learning and collaborating. The site offers tutorials, guides, and documentation for all skill levels. However, one of the most valuable features is the **Arduino Forum**.

- **What You'll Find**:
    - Troubleshooting help
    - Project ideas and inspiration
    - Code snippets and shared libraries
    - Community-driven answers to technical problems
- **Why It's Great**: The forum is organized by topics such as **General Electronics**, **Programming**, and **Projects**. You can search for solutions to specific problems or ask new questions. Many of the answers come from experienced users, including professionals and enthusiasts alike, making this forum a treasure trove of practical knowledge.

You can visit the official Arduino website and forum here: Arduino Forum.

2. Reddit Communities

Reddit is one of the largest and most diverse online platforms for discussion, and it has many subreddits dedicated to Arduino. These communities are great for both beginners and experts looking to engage with others.

- **r/arduino**: One of the largest and most active subreddits, this community features discussions on a variety of topics, from beginner tips to advanced projects. Users share tutorials, show off their projects, and help each other solve problems.
- **r/robotics**: If you're also interested in robotics and not just Arduino, this subreddit provides discussions about everything from robot design to complex algorithms. Many Arduino projects are shared here as well.
- **r/electronics**: While not exclusive to Arduino, this subreddit is ideal for discussing the electronics aspects of your Arduino projects, such as circuits and sensors.
- **r/learnprogramming**: For those who want to focus more on the coding side, this subreddit can offer advice on learning programming and improving your code, which directly applies to Arduino programming.

You can explore these communities here:

- r/arduino
- r/robotics
- r/electronics
- r/learnprogramming

3. Instructables

Instructables is a well-known online platform for DIY enthusiasts and hobbyists to share their creations, including Arduino-based projects. You can find step-by-step guides, videos, and tutorials for a wide range of projects, from beginner to expert level.

- **What You'll Find**:
    - **Project Tutorials**: You'll find clear, detailed tutorials on a wide variety of Arduino projects.
    - **User Interactions**: You can ask questions directly on each tutorial page and receive answers from the community.
    - **Collaboration Opportunities**: Many project creators on Instructables are open to feedback and collaboration, and you can follow projects or leave comments to interact with others.

Instructables fosters a spirit of community, with users actively sharing advice and improvements on each project. It's a great place to both learn and contribute.

Explore Instructables here: Instructables Arduino Projects.

4. GitHub

GitHub is an essential platform for sharing code, collaborating on projects, and contributing to open-source initiatives. Many Arduino developers and enthusiasts use GitHub to host their code, project documentation, and libraries.

- **What You'll Find**:

- **Open-Source Projects**: Many Arduino libraries, projects, and full-fledged systems are available on GitHub. You can explore these projects to learn from real-world applications.
- **Collaborative Projects**: You can contribute to other people's projects, report bugs, or submit new features.
- **Sharing Your Work**: If you've developed an Arduino project, GitHub provides an excellent way to share it with the world and invite others to collaborate.

GitHub is a hub for coders, engineers, and makers. Whether you're looking to collaborate or learn from existing repositories, it's an invaluable resource for expanding your Arduino skills.

Check out Arduino-related projects on GitHub: GitHub Arduino.

5. Facebook Groups and Discord Servers

Facebook and **Discord** host numerous groups and servers dedicated to Arduino. These platforms are great for real-time conversations and quick troubleshooting.

- **Facebook Groups**: Search for groups like **Arduino Projects** or **Arduino Hackers**, where members post their own projects, give advice, and discuss hardware and software issues. These groups are often full of friendly hobbyists who are eager to share their knowledge.
- **Discord Servers**: Discord is a voice and chat platform with many servers dedicated to electronics,

robotics, and Arduino. Servers like **Arduino Hub** or **The Robot Lab** offer channels for different topics, including general advice, project ideas, and coding assistance.

These platforms allow for more informal interactions, real-time collaboration, and problem-solving, making them a perfect addition to your online resources.

6. Hackster.io

Hackster.io is another fantastic community-driven platform for makers, engineers, and programmers. Hackster.io allows users to publish their own projects, ranging from beginner to advanced levels.

- **What You'll Find**:
    - **Project Galleries**: Browse through thousands of user-uploaded projects, including detailed tutorials on using Arduino for various applications.
    - **Hackster Contests**: Participate in themed competitions to win prizes and gain recognition for your work.
    - **Learning Pathways**: Hackster.io offers curated learning paths for those looking to explore Arduino in a structured way.

Hackster.io is a hub for project creators and enthusiasts to learn, share, and collaborate, and it's a great way to expand your skills while contributing to the broader maker community.

Check out Hackster.io here: <u>Hackster.io Arduino Projects</u>.

7. YouTube Communities

While YouTube is primarily a video platform, many creators who specialize in Arduino and robotics actively foster communities in the comments and during live streams. Channels such as **Paul McWhorter's Arduino Tutorial** series, **GreatScott!**, and **Jeremy Blum's Arduino Tutorials** have thriving communities where viewers exchange ideas and troubleshoot issues together.

- **What You'll Find**:
    - **Tutorial Videos**: Clear, step-by-step videos explaining Arduino concepts.
    - **Live Streams and Q&A**: Channels often host live Q&A sessions where you can ask questions and get help in real-time.
    - **Project Sharing**: Many creators encourage viewers to share their own projects in the comments or in dedicated groups.

You can find these channels on YouTube, and by joining the comments section or live discussions, you can interact with other learners.

Chapter 7: Building Smart Projects: From Concept to Creation

As the world of technology evolves, the ability to create **smart projects** is becoming more accessible than ever. With Arduino, you can build a variety of intelligent systems, from smart home devices to sensor-based applications that react to their environment. In this chapter, we will explore how to take a concept—an idea—and turn it into a fully functional, smart project using the power of Arduino. By focusing on project development from **conceptualization** to **creation**, you will learn how to plan, design, and execute a smart system that can solve real-world problems.

The Power of Smart Projects

A "smart project" refers to any system or device that uses technology to gather, process, and react to information in real-time. In the context of Arduino, these systems often involve sensors (for input), actuators (for output), and microcontrollers (for processing data). These elements work together to create an intelligent system that can respond to environmental changes, make decisions based on data, and even learn from its surroundings.

Examples of smart projects include:

- **Smart lighting systems** that adjust brightness based on room occupancy or ambient light.
- **Weather stations** that track temperature, humidity, and pressure in real-time, sending data to your phone or cloud services.

- **Home automation systems** that control appliances, security cameras, and heating based on user preferences or external factors.
- **Robots** that can follow paths, avoid obstacles, or even interact with humans intelligently.

By building such projects, you'll not only learn how to combine various components but also develop skills in systems thinking, problem-solving, and creativity—skills that will serve you well in many areas of technology and engineering.

Step 1: Identifying the Concept

The first step in building any smart project is identifying what you want to build. Before diving into the technicalities, it's important to ask yourself a few key questions:

1. **What problem does my project solve?** Think about everyday issues or inefficiencies that could be improved with technology. For example, does your room have too many switches for lights, fans, or appliances? Could a smart system solve that?
2. **What will my project do?** Consider the functions you want your project to have. Should it sense temperature and automatically adjust the room's climate? Or maybe it could send a notification when something needs attention, like a watering system for your plants?
3. **Who will use this project?** Think about whether the project is for personal use, a group of users, or even for sale. The complexity, interface, and usability should be influenced by who will be interacting with the system.

4. **How will my project improve the user experience?** A smart project should provide tangible benefits to the user. Perhaps it will save time, energy, or money, or offer more control and flexibility than existing alternatives.

Once you've nailed down your project idea, you can begin outlining its components, features, and user interactions.

Step 2: Defining the System Requirements

A great way to keep your project on track is to clearly define the system's requirements before you begin the build. This ensures that you understand what needs to be done and provides a clear roadmap for success. Here are some things to consider:

- **Hardware Requirements**: What physical components will you need? These might include:
    - **Arduino board** (e.g., Arduino Uno, Nano, or Mega)
    - **Sensors** (e.g., temperature, light, humidity, motion)
    - **Actuators** (e.g., motors, LEDs, servos, relays)
    - **Displays** (e.g., LCD, OLED screens)
    - **Power supply** (e.g., battery, USB power, or external adapter)
    - **Connectivity modules** (e.g., Bluetooth, Wi-Fi, Ethernet)
- **Software Requirements**: What will your system need in terms of coding and software? This could involve:
    - **Arduino IDE** for writing and uploading the code

- Libraries for specific sensors and modules (e.g., DHT11 for temperature and humidity sensors, ESP8266 for Wi-Fi)
    - **Cloud platforms** for storing or accessing data remotely (e.g., ThingSpeak, Blynk, or Firebase)
    - Mobile or web interfaces for interacting with your system
- **User Interface**: If your project will involve a user interacting with it (e.g., controlling a device through a smartphone), think about:
    - What kind of interface will the user experience (buttons, sliders, app notifications)?
    - How will the system react to the user's inputs?

Once the requirements are defined, create a **project flow** diagram that outlines how the system will work from start to finish. This helps visualize data flows and interactions, and serves as a blueprint for later steps.

Step 3: Designing the System

After defining the requirements, the next step is designing your project. This process involves combining your ideas with technical specifications, choosing the right components, and laying out how the system will be built. Here's a step-by-step approach:

1. **Create a Schematic**: A schematic diagram shows how your sensors, actuators, and Arduino board are connected together. This is crucial for understanding the wiring and ensuring that all components work together. You can use tools like **Fritzing** or **Tinkercad** to create and visualize your schematics.

2. **Choose Components Wisely**: Based on your project's requirements, select components that fit your needs. Make sure that the sensors are compatible with your Arduino board, and check the power requirements of each component.
3. **Write Pseudocode**: Before jumping into the actual code, writing pseudocode (a step-by-step, human-readable outline of your program) helps clarify the logic behind your project. Pseudocode helps you plan the structure of your code and can make the coding process smoother.

Step 4: Building the Hardware

Now it's time to physically build the project! Start by setting up the **Arduino board** and connecting your **sensors and actuators** based on the schematic you created. For example, if you're building a smart temperature control system, you'd connect the temperature sensor (e.g., a DHT11 or DHT22) to the Arduino and connect a fan or heater as the actuator.

- **Breadboard Setup**: Use a breadboard to prototype your circuits. This allows for easy changes and adjustments without soldering.
- **Testing Each Component**: Before moving forward, test each sensor and actuator individually with simple code examples. This ensures they work properly before integrating them into the larger project.

Step 5: Writing the Code

With the hardware in place, it's time to bring your project to life through code. This is where you'll translate your pseudocode into functional Arduino code. Start by setting up your Arduino IDE, importing the necessary libraries for your sensors and actuators, and structuring the program flow. Here are some key considerations when writing your code:

1. **Reading Inputs**: Use the correct functions to read data from your sensors. For example, for a temperature sensor like the DHT11, you would use the read() function.
2. **Processing Data**: You can add conditions to process the sensor data, like triggering a response when the temperature crosses a certain threshold.
3. **Control Outputs**: Use digitalWrite() and analogWrite() to control the actuators based on the sensor data. For example, if the temperature exceeds 30°C, turn on the fan.
4. **Debugging and Testing**: Test your code in stages. If you run into errors, troubleshoot them by checking the connections, reviewing the code for syntax issues, and using Serial.println() to monitor sensor readings and outputs.

Step 6: Testing and Refining the Project

After coding and connecting everything, the real fun begins: testing. Run your project and observe how it behaves. Does it respond correctly to inputs? Does it perform as expected? At this stage, it's essential to fine-tune your project:

- **Adjust Sensitivity**: Maybe the temperature sensor isn't detecting minor fluctuations, or the motion sensor is too sensitive. Make adjustments in both the hardware (e.g., moving the sensor) and software (e.g., modifying the thresholds).

- **Optimize Code**: Ensure that the code runs efficiently. For more complex projects, consider optimizing your code to reduce the Arduino's workload.
- **Add Extra Features**: As you test, think of ways to enhance your project. Could it have a mobile app to receive notifications? Could you add more sensors for more advanced functionality?

Step 7: Finalizing Your Smart Project

After you've perfected your project, it's time to finalize it. Consider transitioning from a **breadboard setup** to a **permanent circuit** by soldering components onto a **PCB (Printed Circuit Board)** for a cleaner and more durable setup. You may also want to build an **enclosure** to protect your components, especially if the project is intended for long-term use.

- **Power Supply**: Ensure your power supply is stable. If your project requires a battery, select one with the right voltage and capacity.
- **User Interface**: If your project involves a user interface, make it as intuitive as possible. Consider adding an LCD screen for feedback or developing a mobile app for remote control.

Guide to Prototyping: Turning Your Idea into a Working Arduino Project

Prototyping is the key process through which an idea transitions from a conceptual stage into a functional reality. Whether you're building a smart home device, an automated garden system, or a simple robot, the principles of prototyping remain the same. It is a process of exploration, experimentation, and iteration, where you transform a basic idea into a working model. In this guide, we'll walk you through a detailed example of how to prototype a **smart home device**—specifically, a **smart light control system**—using Arduino. By the end of this chapter, you'll have the tools and insights to prototype any idea, from smart home devices to more complex automation systems.

Step 1: Define the Idea and Concept

Before you can begin prototyping, it's crucial to define your idea clearly. Prototyping works best when you have a vision of what you want to achieve and can break it down into manageable steps. Let's take the example of a **smart light control system** that automatically turns lights on or off based on room occupancy. The basic functionality we want is:

1. **Sensor Detection**: Detect whether a person is in the room using a motion sensor.
2. **Control Lights**: Turn the lights on if motion is detected and turn them off after a set period if no motion is detected.
3. **User Input**: Allow users to manually control the lights through a button or mobile app.

With the concept in mind, the next step is to outline how the system will function. Break down the problem into its components:

- **Sensors**: You'll need a motion sensor (PIR sensor) to detect movement in the room.
- **Actuators**: Use a relay or transistor to control the light's power.

- **Microcontroller**: An Arduino board (e.g., Arduino Uno) will process the sensor's input and control the light.
- **User Interface**: Optionally, a button or a mobile app can allow the user to override the system.

Step 2: Define Requirements and Specifications

Now that you have a basic idea of how the system will function, you need to specify the **hardware** and **software** requirements for the project. Here's a detailed breakdown:

Hardware Components:

1. **Arduino Uno**: This microcontroller will process inputs and control outputs.
2. **PIR Motion Sensor**: Detects motion in the room to trigger the light.
3. **Relay Module**: Used to control the high-voltage light circuit via Arduino.
4. **Button**: Allows users to manually control the light.
5. **LED Light (optional)**: A simple LED light for testing purposes.
6. **Jumper wires**: To make connections between components.
7. **Power Supply**: For the Arduino and relay module.

Software Components:

- **Arduino IDE**: For writing and uploading the code to the Arduino board.
- **Libraries**: You might need libraries such as Adafruit_Motion for motion sensors, or a simple Relay library to control the relay.

- **User Interface Code**: If you add buttons or app-based control, you'll write code to manage user input.

Functional Requirements:

- **Motion Detection**: The motion sensor should detect movement and trigger the lights to turn on.
- **Auto Off**: After a specified time (e.g., 5 minutes) of no motion, the lights should turn off automatically.
- **Manual Override**: A button should allow the user to override the automatic system and control the light manually.

Step 3: Design the Circuit

Now that you have your hardware and software specifications ready, it's time to design the circuit. Start with a basic schematic and layout to visualize how the components will connect:

1. **Connect the PIR sensor**: Connect the VCC and GND pins of the PIR sensor to the Arduino's 5V and GND pins. The signal pin will go to one of the Arduino's digital input pins (e.g., pin 7).
2. **Connect the Relay Module**: Connect the relay's VCC and GND to the 5V and GND on the Arduino. The relay's control pin will connect to a digital output pin (e.g., pin 8).
3. **Connect the Button**: The button will be connected to one of the digital pins (e.g., pin 9) and should be configured as a **pull-down resistor** so the button sends a high signal when pressed.
4. **Power the Light**: If you're using an LED light, you can power it directly through the relay. For high-

voltage systems (like actual room lights), you'll need a relay capable of handling higher current.

Explanation of the Code:

- **Motion Detection**: The code reads the motion sensor every loop iteration. If motion is detected, it turns the light on. If no motion is detected for 5 minutes, the light is automatically turned off.
- **Manual Override**: A button allows users to toggle the light on and off manually, overriding the motion-based control.
- **State Management**: The lightState variable keeps track of whether the light is on or off, ensuring the system behaves as expected.

Step 5: Build and Test the Prototype

With the code written and the circuit built, it's time to bring everything together and test the system. Here's a step-by-step approach to testing the smart light control system:

1. **Connect the Arduino** to your computer and upload the code using the Arduino IDE.
2. **Test Motion Sensor**: Walk in front of the motion sensor and ensure the light turns on when motion is detected and turns off after 5 minutes of no motion.
3. **Test Manual Control**: Press the button and verify that it turns the light on and off regardless of motion.
4. **Observe Behavior**: Watch for any bugs or unexpected behavior. For example, if the light doesn't turn off after the time delay, you may need to adjust the logic or timing parameters.

Step 6: Refine and Improve

Once your prototype is working, it's time to refine and improve it:

- **Fine-Tuning**: Adjust the motion sensor's sensitivity or change the time delay before the light turns off.
- **Add Features**: Consider adding more features, such as controlling multiple lights or integrating a mobile app via Bluetooth or Wi-Fi.
- **Optimize Code**: Refactor your code for efficiency, reduce unnecessary delays, or add more sensors for better accuracy.

Product Design Thinking: Approaching Your Arduino Projects with a Creative, Iterative Process

Design thinking is a creative approach to problem-solving that focuses on understanding the needs of the user, brainstorming solutions, creating prototypes, and continuously refining those ideas. In the world of Arduino projects, this mindset is crucial because, as a maker, you're often building innovative solutions from scratch. Arduino allows you to quickly prototype ideas, but without a structured approach to design, your projects could quickly become confusing or dysfunctional.

In this section, we'll explore how to incorporate **design thinking** into your Arduino projects. We'll go through each phase of the process—**Empathy, Define, Ideate, Prototype, Test, and Iterate**—and explain how you can apply these principles to your own projects. By the end of this chapter, you'll be equipped with a methodology to guide your thinking, troubleshoot effectively, and iteratively refine your projects, leading to more functional and creative outcomes.

1. Empathy: Understanding the Problem You're Trying to Solve

The first stage of product design thinking is **empathy**—understanding the problem, context, and needs of your project. In the context of Arduino, this means asking, "What problem does this project solve?" or "What need am I fulfilling?"

Questions to ask yourself:

- Who will use my project? (e.g., Is it for personal use, or is it for a larger audience?)
- What challenge does my project aim to solve? (e.g., A smart garden system to automate watering, or a motion-activated security light)
- What do I want the end-user experience to be like? (e.g., easy, hands-off, or highly interactive)

For example, if you're working on a **smart weather station**, consider:

- How will the data be used by the person? Will it be used to make decisions on planting crops, adjusting room temperature, or tracking weather patterns?
- What type of user interface will make this information accessible and useful to them?

Empathy in Arduino projects goes beyond just understanding the functionality; it's about considering the user experience, comfort, and usability. Make sure the solution you're designing matches the needs of the people who will interact with it.

2. Define: Narrowing Down the Problem

Once you've empathized with the problem, the next step is to **define** it. This phase is about focusing your scope. You'll synthesize the information you've gathered, clearly state the problem, and outline the parameters of your project.

Steps to define your project:

- **Narrow the scope:** Start by identifying the core objectives. Don't overcomplicate things at this stage. For instance, if you're building a **smart home light system**, your primary goal might be to automate the lighting in a specific room based on motion detection.
- **List out requirements:** Write down what your project needs to achieve. This could include controlling devices based on sensor input, using a button for manual override, or ensuring that the system works with multiple sensors.
- **Constraints:** Identify the limitations you'll face, such as power constraints, limited components, budget, time, or technical challenges.

For example, for your **motion-sensing light system**:

- Objective: Automatically turn on a light when motion is detected, and turn it off after 5 minutes of inactivity.
- Constraints: It must be easy to install with minimal wiring, and it should work with standard household bulbs.

By clearly defining the problem, you're establishing a solid foundation for the next phases of the design process. You'll know exactly what you need your project to do, and any extraneous features can be left out or saved for future iterations.

3. Ideate: Brainstorming Possible Solutions

The **ideate** phase is where the magic happens. This is where you brainstorm potential solutions to the defined problem. Your goal during this phase is to generate as many ideas as possible, without worrying about whether they will work or not. Let your creativity run wild!

Steps for ideation:

- **Sketch ideas:** Draw out the system or product you're designing. Use rough sketches, diagrams, or flowcharts to visualize your project.
- **Brainstorm alternatives:** Think about the different ways your problem could be solved. For instance, instead of a motion sensor, could you use a light sensor? Or, could you create a hybrid system where both a light sensor and a motion sensor are used for higher accuracy?
- **Evaluate practicality:** Once you've brainstormed, evaluate each idea in terms of feasibility. What parts do you need? How much time will it take? Do you have the necessary skills?

For your **smart light control system**, here are a few ideas you might consider:

1. **Motion Sensing with Light Adjustment**: Using a motion sensor, the light could also adjust brightness based on time of day.
2. **Voice-Controlled System**: Implementing Bluetooth and voice commands (via a phone or smart speaker) for more flexibility.
3. **App-Based Control**: Allowing users to control the lights remotely via a smartphone app.

At this stage, don't worry too much about the practicality of each idea. The goal is to explore all possibilities and think about every angle.

4. Prototype: Build Your First Model

In the **prototype** phase, you start building physical versions of your ideas. The goal is to quickly and cheaply build a model of your design so you can test its functionality.

Steps for prototyping:

- **Start small:** Begin with a simple version of your project. Don't worry about making it perfect—just focus on creating a working prototype that demonstrates your idea.
- **Use existing parts:** Leverage Arduino components that you already have, such as sensors, motors, and relays, to prototype quickly.
- **Focus on functionality:** Ensure your prototype can execute the core functions of your project. For a **smart light system**, this means getting the motion sensor to trigger the light on/off, even if the design is messy or unpolished.

For example, your prototype might include:

- A **PIR sensor** connected to the Arduino.
- A **relay module** that turns a light on/off when activated.
- A basic **LED** (as a stand-in for an actual light) to test the system.

Keep your prototype basic and focused solely on proving the concept.

5. Test: Evaluate and Analyze

Once your prototype is built, the next step is to **test** it. In this phase, you'll evaluate the functionality of your design, note any issues, and gather feedback. You may encounter technical issues or user experience problems that were not evident during the prototyping phase.

Testing methods:

- **Functional testing**: Does the system work as expected? Does the light turn on when motion is detected, and does it turn off after the set delay?
- **User testing**: Have others use your system. Do they find the light control intuitive? Are there any usability issues (like button placement or sensor placement)?
- **Stress testing**: Does your system perform well under varying conditions (e.g., different lighting, high motion activity, etc.)?

6. Iterate: Refine and Improve

After testing your prototype, you'll likely encounter problems or areas for improvement. The **iteration** phase is where you refine and improve your design based on the feedback and observations from the testing phase.

Steps for iteration:

- **Identify issues:** What didn't work? Was the sensor too sensitive? Did the light stay on too long? Was the relay too slow to respond?
- **Make improvements:** Adjust your design or code. For example, you could change the motion sensor's sensitivity, extend the time before the light turns off, or add additional sensors for better accuracy.
- **Re-test:** After making changes, go back to the testing phase to ensure the improvements worked.

Design thinking is a cyclical process, and iteration is key to perfecting your project. With each cycle, your project will improve, becoming more functional, reliable, and user-friendly.

Creative Freedom: Unleashing Your Imagination in Arduino Projects

One of the most exciting aspects of working with Arduino is the creative freedom it offers. As you learn the foundational skills and gain confidence in building basic circuits and coding, the next step is to harness your creativity and bring your personal ideas to life. Arduino is not just about following step-by-step instructions; it's a platform that encourages experimentation, customization, and innovation. It empowers you to design your own projects, solve real-world problems, and explore new concepts.

In this chapter, we'll delve into how you can embrace your creative freedom and turn your ideas into tangible, functional creations. We'll provide suggestions on how to customize your projects based on your interests and passions, and give you the tools and techniques to take full ownership of your projects. Whether you're interested in art, music, technology, or solving everyday challenges, Arduino gives you the flexibility to explore all these areas and more.

1. Let Your Interests Guide Your Projects

One of the best ways to spark your creativity is to align your Arduino projects with your personal interests. Arduino projects are only limited by your imagination, and by connecting your projects to something you care about, you make the learning process more fun, rewarding, and meaningful.

Here are some ways to integrate your personal interests:

- **Music and Sound:** If you're passionate about music, consider building an **Arduino-powered synthesizer**, a **sound-reactive light system**, or even a **digital drum kit**. By integrating sound sensors, you can create projects that interact with music, allowing for visual and audio experiences that respond to the rhythm or frequency of sound.
- **Smart Home Devices:** If you're interested in automation and convenience, Arduino offers an excellent gateway to building your own **smart home devices**. You could create custom systems such as a **smart thermostat**, **motion-activated lights**, or even a **voice-controlled fan** using Arduino and various sensors.

- **Environmental Projects:** For those who are environmentally conscious, Arduino can help create sustainable solutions. You could build a **solar-powered garden irrigation system**, a **temperature and humidity monitoring station**, or a **recycling bin with automatic sorting**. These projects not only serve practical purposes but also help in promoting sustainability and conservation.
- **Art and Design:** If you have a creative side that leans toward art, try integrating Arduino with your artistic projects. Build **interactive sculptures**, **kinetic art**, or **light installations** that change based on sensor input or user interaction. With Arduino, your art can come alive, adding a whole new layer of interactivity and creativity.
- **Robotics:** For robotics enthusiasts, Arduino is the ideal platform to start building your own robots. Create a **line-following robot**, a **robotic arm**, or even a **drone**. Arduino offers the flexibility to experiment with sensors, motors, and control systems, making it a perfect entry point into the world of robotics.
- **Health and Wellness:** For those passionate about health tech, Arduino can help create projects like a **fitness tracker**, a **heart rate monitor**, or a **sleep analysis system**. With the right sensors, you can track vital signs and use that data to build tools that improve personal health and wellness.

2. Customizing Projects: Adding Your Personal Touch

Once you have a basic understanding of Arduino, the next step is to think about customization. Customization allows you to tweak your projects to fit your specific needs or preferences, giving them a more personal touch. You can experiment with different components, adjust the functionality, or integrate unique features based on your vision.

Here's how to get started with customizing your projects:

- **Sensors and Actuators:** The variety of sensors and actuators available for Arduino gives you endless customization options. You can modify the type of sensors used based on the desired outcome. For example, if you're building a weather station, you can choose different types of temperature sensors—**DHT11** for basic projects or the more precise **DS18B20** for accurate readings.
- **Display and User Interface:** You can personalize the way information is presented to users by experimenting with different displays like **LCD**, **OLED**, or even **touchscreens**. If you're building a smart thermostat, you could customize the interface to show the temperature, humidity, and even control settings through a touchscreen or a web interface.
- **Advanced Features:** Arduino allows you to take your basic projects to the next level with advanced features. If you've built a simple motion-sensing light system, why not add a **time-of-day feature**? The light could stay on for longer at night and turn off quickly during the day. If you're into robotics, try adding **machine learning algorithms** to allow your robot to learn and improve over time.
- **Mobile Integration:** Many projects can be expanded to include **mobile app integration** using **Bluetooth** or **Wi-Fi**. This opens the door to making your projects

more interactive. Imagine controlling a robotic arm from your smartphone or adjusting your home's lighting and temperature settings while on the go.
- **Aesthetic Customization:** Arduino projects don't just have to be functional—they can also be visually appealing. Consider how you can make your project stand out by using custom enclosures, colorful LEDs, or 3D-printed parts. Customizing the appearance of your projects not only makes them more fun to build but also allows you to showcase your unique style.

3. Overcoming Challenges and Embracing Failure

When it comes to creativity, it's important to remember that the process of experimentation often involves failure. The beauty of working with Arduino is that it allows for rapid prototyping and testing. If something doesn't work as expected, you can always make adjustments and try again.

Here are some tips to embrace failure as part of the creative process:

- **Fail Forward:** Every failure is an opportunity to learn. If your motion sensor doesn't trigger your light, troubleshoot the wiring, check the code, or adjust the sensor's sensitivity. With each failure, you'll become more skilled at problem-solving and critical thinking.
- **Iterate and Improve:** Don't expect your project to be perfect right away. Start with a basic prototype, test it, and refine it over time. Iteration is a key part of the creative process, and it's how you'll end up with something truly unique and functional.
- **Seek Inspiration, Not Imitation:** While it's tempting to copy other people's projects, the real fun comes

from adding your unique twist. Once you've built the core of a project, think about how you can modify it to reflect your personal style or needs. This is where true creativity shines—taking an existing idea and making it your own.

4. Collaboration and Sharing Ideas

As you continue on your Arduino journey, remember that creativity thrives in a community. Don't be afraid to share your ideas and projects with others. Collaborating with fellow makers can provide valuable feedback, and learning from others' experiences can inspire new directions for your own projects.

Here's how to get involved with the Arduino community:

- **Online Communities:** Join online forums like **Arduino's official community**, **Instructables**, or **Reddit's Arduino** to find projects, share ideas, and ask questions. You'll find that the Arduino community is incredibly supportive and full of creative thinkers who are always willing to help.
- **Local Meetups and Hackathons:** If you're looking for a more hands-on experience, consider attending local maker meetups or hackathons. These events often provide an excellent opportunity to work on collaborative projects, share knowledge, and gain fresh ideas for your next big Arduino project.
- **Open-Source Collaboration:** Arduino projects are often open-source, meaning that you can contribute to existing projects or reuse others' code and ideas to accelerate your own designs. Whether it's improving a piece of software, adding a new feature, or

contributing hardware designs, open-source collaboration is a great way to both learn and contribute to the community.

5. Inspiring Others and Becoming a Maker Leader

As you become more proficient in Arduino, you may want to inspire others to start their own creative journeys. Share your projects, mentor beginners, or start a blog or YouTube channel to teach others. By teaching, you deepen your own understanding and become part of a larger movement of makers who are pushing the boundaries of innovation.

Your personal growth as a maker can also inspire the next generation of inventors. Your projects might spark curiosity in others, leading them to create their own Arduino-based solutions.

Unlocking Infinite Possibilities

Arduino is a platform that thrives on creativity, and with every project you build, the potential for innovation grows. The beauty of Arduino is that it allows you to take your wildest ideas, test them, refine them, and eventually bring them to life. Whether you're exploring robotics, building a smart home device, or experimenting with interactive art, Arduino gives you the tools to create something truly unique.

Remember, the sky's the limit. Keep experimenting, customizing, and most importantly, have fun! Your next great idea could be just one circuit away.

Chapter 8: Advanced Arduino Projects: Pushing the Limits of Your Skills

As you continue your journey with Arduino, you'll find that the more you learn, the more you want to challenge yourself. This chapter is all about pushing the boundaries of what you've learned and testing your skills with advanced projects that integrate multiple components, sensors, and complex coding techniques. These projects are designed to help you hone your problem-solving abilities, deepen your understanding of electronics, and encourage you to think outside the box.

1. Why Push Your Limits?

When you start with Arduino, you learn the basics—how to wire up simple circuits, write basic code, and interface with a few sensors and actuators. These initial projects give you the foundation needed to understand how Arduino works. But as you get comfortable with these basics, you'll naturally want to explore more challenging projects that stretch your imagination and abilities.

Pushing the limits of your skills allows you to:

- **Expand Your Knowledge:** Advanced projects require integrating more complex components and sensors, which means you'll be learning new concepts, like communication protocols, data processing, and advanced control algorithms.
- **Gain Real-World Experience:** By working on complex projects, you'll not only increase your

technical proficiency but also gain hands-on experience with systems that are used in real-world applications.
- **Solve Bigger Problems:** As you build more advanced projects, you'll begin solving problems that require creativity and more sophisticated solutions. This empowers you to approach challenges with a more robust toolkit.

2. Integrating Multiple Components: The Key to Complexity

One of the hallmarks of advanced Arduino projects is the integration of multiple sensors, actuators, and systems. Rather than working with one sensor and one actuator at a time, advanced projects require you to combine several components into a cohesive system.

Example Projects for Integration:

- **Smart Home Automation System:** This project integrates temperature sensors, motion detectors, and light actuators to automate home appliances. Using an Arduino board, you could create a system where lights turn on when motion is detected, and the thermostat adjusts the temperature based on the time of day or occupancy.
- **Weather Station:** Combining temperature, humidity, and pressure sensors, this project will give you the ability to gather environmental data and display it on an LCD screen. You'll learn how to work with real-time data collection, and possibly store the data in a cloud database for remote access.

- **Smart Security System:** For this project, you could integrate motion sensors, cameras, and door locks. When motion is detected, the system sends a notification to your phone or triggers a camera to take a picture, which can be viewed in real-time through a web interface.
- **Robot Arm with Feedback Control:** This is an advanced robotics project that integrates motors, encoders, and sensors to create a robot arm that responds to feedback. You'll need to write more sophisticated code, incorporating principles of **PID (Proportional-Integral-Derivative) control** to ensure smooth, responsive movements.

These projects require more than just basic wiring—they require an understanding of how systems work together. Arduino allows you to create interdependent systems where each component plays a crucial role in achieving the desired functionality.

3. Advanced Coding Techniques: Moving Beyond the Basics

While the fundamental programming concepts in Arduino—such as variables, loops, and conditional statements—are essential, advanced projects push you to explore more advanced coding techniques that can help you manage complexity and optimize your systems.

Key Coding Concepts for Advanced Projects:

- **Libraries and External Code:** As you work on more complex systems, you'll often find that libraries can help simplify your project. For example, instead of

writing code from scratch to control a motor, you can use a pre-existing library that provides easy-to-use functions. Learning to leverage external libraries will save time and make your code cleaner.
- **Interrupts:** Interrupts allow the Arduino to react to events without constantly checking for them in the main loop. For example, you can set an interrupt to trigger when a button is pressed, allowing your system to react immediately to external events without interrupting ongoing processes.
- **State Machines:** A state machine is a powerful programming concept that allows you to break your program into distinct states. This is particularly useful for managing complex projects like robots, where the system might need to switch between modes—e.g., idle, moving, or detecting an obstacle.
- **Communication Protocols:** Many advanced projects involve communication between devices. Common communication protocols used in Arduino include **I2C**, **SPI**, and **UART**. These protocols allow multiple devices to communicate with each other, expanding your project's capabilities.
- **Wireless Communication:** Many advanced Arduino projects use wireless modules like **Bluetooth, Wi-Fi (ESP8266/ESP32),** or **Zigbee** to enable communication between multiple Arduino boards or between an Arduino and a smartphone or computer. This opens up the world of IoT (Internet of Things) applications.
- **Machine Learning and Artificial Intelligence:** With advanced sensors and computational power, Arduino is capable of integrating machine learning algorithms. By training a model (e.g., using a computer) and uploading it to Arduino, you can make decisions

based on sensor inputs, like predicting sensor behavior or identifying patterns in the environment.

4. Advanced Robotics: Moving From Theory to Functionality

Robotics is one of the most exciting areas of advanced Arduino projects, combining coding, electronics, and mechanical design. As you move beyond basic robots, you'll start integrating more sensors and advanced control algorithms to create robots that can interact intelligently with the world around them.

Advanced Robotics Projects:

- **Autonomous Vehicles:** Building a fully autonomous robot or car that can navigate a maze, avoid obstacles, or follow a pre-set path is a great way to learn advanced robotics. You'll incorporate sensors like ultrasonic rangefinders and infrared sensors for navigation, and use algorithms like **PID control** for movement precision.
- **Robotic Arm with Servo Motors and Feedback:** To take your robotic arm to the next level, you can implement feedback control using sensors (like encoders) to ensure accurate positioning. This could involve complex inverse kinematics algorithms to control the arm's movements with precision.
- **Voice-Controlled Robots:** By integrating a **voice recognition module** or connecting to a cloud service, you could create a voice-controlled robot. This involves understanding how speech processing works and how to convert that data into actionable commands for controlling motors or sensors.

5. Troubleshooting and Optimizing Your Advanced Projects

As your projects grow in complexity, troubleshooting and optimization become crucial skills. Here's how to tackle challenges as you work on advanced projects:

1. Debugging Techniques:

- **Serial Monitor:** One of the best tools for debugging is the **Serial Monitor**. This feature allows you to print values from your sensors or variables directly to the computer to monitor how your system is behaving in real-time. By printing critical data points, you can quickly identify where things go wrong in your code or circuit.
- **Multimeter and Oscilloscope:** A **multimeter** is invaluable for checking whether your sensors and components are receiving the right voltage. An **oscilloscope** is even more powerful, allowing you to visualize signals and ensure your components are communicating correctly.
- **Modular Testing:** As you add new components to your system, test them individually first to make sure they're functioning properly. This can prevent cascading errors and help you pinpoint issues more easily.

2. Optimizing Performance:

- **Code Optimization:** As you move toward more complex projects, you may encounter performance bottlenecks, especially if you're using an older or less powerful Arduino model. In this case, optimizing your

code to use fewer resources (memory, processing power) will improve the performance of your project.
- **Power Consumption:** When you integrate multiple sensors and actuators into your project, power management becomes crucial. Consider using **low-power sensors** and employing **sleep modes** to conserve energy, especially for battery-operated systems.
- **Modular Design:** Consider breaking down your projects into smaller, more manageable subsystems. This will not only make troubleshooting easier but also allow you to improve individual components without overhauling the entire system.

6. Real-World Application: Translating Skills to Industry

Once you've mastered advanced Arduino projects, the skills you've gained can be applied to real-world engineering problems. Whether you're interested in working on industrial automation, IoT solutions, or cutting-edge robotics, the foundations you've built through Arduino will serve as a stepping stone for more advanced applications in the professional world.

Career-Ready Skills:

- **Embedded Systems Development:** Arduino is often used as a tool for teaching embedded systems development, which is a core component of hardware and software engineering.
- **Internet of Things (IoT):** By working with Arduino's Wi-Fi and Bluetooth capabilities, you can develop IoT devices that monitor and control devices remotely.

- **Prototyping for Startups:** Many tech startups use Arduino to prototype new products before committing to expensive manufacturing processes. Understanding how to create functional prototypes can give you a competitive edge in the job market or as an entrepreneur.

Challenge the Reader: Intermediate Projects to Expand Your Skills

Now that you have a solid understanding of basic Arduino programming and have built simple projects, it's time to take your skills to the next level. The real excitement in Arduino lies in the ability to design and build more complex systems that solve real-world problems. In this section, we'll challenge you with some intermediate projects that will push your creativity, expand your technical knowledge, and help you master important skills like wireless communication, automation, and data handling.

These projects will not only test your existing knowledge but also help you learn new concepts such as Bluetooth control, home automation, and IoT (Internet of Things) systems—each of which is highly applicable in today's tech-driven world.

1. Bluetooth-Controlled Robot

One of the most exciting ways to challenge yourself is by building a Bluetooth-controlled robot. This project involves both hardware and software, and it will require you to get comfortable with wireless communication and motor control. The goal is to create a small robot that can be controlled remotely using a smartphone or tablet via Bluetooth.

Key Concepts Covered:

- **Bluetooth Communication:** Learn how to use the **HC-05 Bluetooth module** to wirelessly communicate with your Arduino board.
- **Motor Control:** Use **motor drivers** to control the direction and speed of the robot's motors.
- **Mobile App Interface:** You'll need to create or use an existing mobile app that sends commands to your Arduino through Bluetooth.

How It Works:

In this project, your Arduino will receive commands via Bluetooth, which will tell it to move forward, backward, left, or right. The Bluetooth module on the Arduino will communicate with a smartphone app, and based on the input (e.g., a joystick on the app), the robot will perform the corresponding action. This project introduces you to the world of remote control systems and lays the foundation for building more complex wireless projects in the future.

Next Steps:

Once you've mastered the basics of the Bluetooth-controlled robot, you can enhance it by adding more features:

- **Add sensors** for obstacle avoidance (e.g., ultrasonic sensor).
- Integrate a **camera module** for real-time video streaming.
- Implement more advanced control algorithms, like **PID control** for smooth navigation.

2. Home Automation System

Home automation is a rapidly growing field, and building your own system with Arduino will help you understand how the Internet of Things (IoT) can be applied to everyday life. A home automation system can automate various tasks in your home, like controlling lights, fans, thermostats, and more—all from a smartphone app or voice assistant.

Key Concepts Covered:

- **Relay Control:** Learn how to control high-voltage appliances (like lights) using **relays** that interface with the Arduino.
- **Smartphone App Integration:** Connect the Arduino to your smartphone using **Bluetooth** or **Wi-Fi** (using ESP8266 or ESP32).
- **Automation Logic:** Program the Arduino to control devices based on time, sensor input, or remote commands.

How It Works:

A basic home automation system could start by controlling a light or fan using a mobile app. When you send a signal from your phone, the Arduino will turn the relay on or off, thus controlling the device connected to it. From here, you can expand the system to manage multiple devices in your home, such as turning on a coffee maker in the morning or controlling your HVAC system.

You could also integrate sensors to make the system more intelligent. For example:

- **Motion sensors** can trigger lights to turn on when someone enters the room.
- **Temperature sensors** could adjust the thermostat or fan based on the room's temperature.
- **Voice assistants** (like Alexa or Google Assistant) can be integrated for hands-free control.

Next Steps:

Once you've successfully built a basic home automation system, you can continue enhancing it by:

- Adding **data logging** to monitor and track energy usage.
- Integrating **voice control** using a **Wi-Fi module** or **Alexa** integration.
- Expanding to multiple rooms or even controlling appliances from anywhere using cloud-based IoT platforms.

3. IoT Weather Station

Building an **IoT weather station** is an excellent intermediate project that will teach you how to collect and analyze data from sensors, and how to send that data over the internet for remote monitoring. This is a great way to explore the power of Arduino in IoT applications, where real-time data is collected and shared.

Key Concepts Covered:

- **Sensors:** Learn how to interface with temperature, humidity, and pressure sensors (such as the **DHT11**, **BMP180**, or **BME280**) to collect environmental data.

- **Wi-Fi Communication:** Use the **ESP8266** or **ESP32** to connect your Arduino project to the internet, enabling remote access and monitoring.
- **Cloud Platforms:** Send your data to cloud platforms like **ThingSpeak** or **Adafruit IO** for data visualization and analysis.

How It Works:

The weather station uses sensors to gather environmental data, such as temperature, humidity, and pressure. This data is then sent wirelessly via Wi-Fi to an IoT platform (like **ThingSpeak**), where you can view real-time graphs and analyze trends. The cloud platform stores your data, so you can access it from anywhere and even set up alerts or notifications if certain conditions are met (e.g., if the temperature exceeds a certain threshold).

Next Steps:

Once you've mastered the basic weather station, you can expand it by:

- Adding **more sensors** like **rainfall**, **wind speed**, or **UV index**.
- **Automating responses**: For example, turn on a fan when the temperature exceeds a set value or trigger an alarm when a sudden change in humidity is detected.
- Integrating a **local display** to show live data, using an **OLED** or **LCD screen**.

4. Smart Plant Watering System

For a more practical and useful project, the **smart plant watering system** allows you to monitor the moisture level of your plant's soil and automatically water it when it gets too dry. This project combines environmental sensors with automation to create a system that ensures your plants are well taken care of, even when you're not around.

Key Concepts Covered:

- **Soil Moisture Sensors:** Learn how to interface with sensors like the **capacitive moisture sensor** to monitor soil moisture levels.
- **Relay Control:** Use a relay to control a water pump or solenoid valve that releases water into the plant.
- **Automation and Timing:** Program the Arduino to water the plant based on predefined moisture thresholds or set times.

How It Works:

The system continuously measures the moisture level in the soil. When the moisture level drops below a set threshold, the Arduino sends a signal to activate the water pump, which then waters the plant. You can program the system to water the plant at regular intervals, ensuring that the plant never goes too long without water.

Next Steps:

To improve the smart watering system, you could:

- Add a **real-time clock** (RTC) module to water the plants at specific times of day.

- **Connect it to the cloud** to monitor the soil moisture remotely and receive alerts if water levels get too low.
- Implement an **automatic shut-off** if the system detects excessive water usage, preventing over-watering.

5. Smart Door Lock System

The **smart door lock** system allows you to lock or unlock your door remotely using Bluetooth or Wi-Fi. This is a great project to dive into security systems and learn about actuators, sensors, and secure communications.

Key Concepts Covered:

- **Servo Motors:** Learn how to control a **servo motor** to physically lock or unlock a door.
- **Wireless Control:** Use Bluetooth or Wi-Fi to send commands to the Arduino from your smartphone or another device.
- **Security Features:** Add security features like password protection, fingerprint scanning, or an RFID key fob for extra layers of access control.

How It Works:

In this project, you will use a servo motor to physically lock or unlock a door when a command is sent via Bluetooth or Wi-Fi. You can control the lock using a mobile app, or implement a more secure method like a **RFID reader** or a **keypad** to provide access.

Next Steps:

You could enhance the system by adding:

- **Multiple layers of security** (fingerprint scanning, RFID, keypad).
- **Notification system**: Send notifications to your phone when the door is unlocked, or if there is an unauthorized attempt to unlock it.
- Integrating with **cloud services** for remote control.

Ready to Tackle the Next Level

These intermediate projects will expand your knowledge and help you master key concepts like wireless communication, automation, and data handling. Each project encourages problem-solving, creativity, and experimentation, and will provide you with practical experience that can be applied to real-world applications.

Explain Advanced Concepts: Unlocking the Full Potential of Arduino

Now that you've gained a solid understanding of Arduino basics, it's time to delve into more advanced concepts that will open up new possibilities for your projects. Understanding these advanced techniques will enable you to take your creations to the next level, allowing you to build more efficient, interactive, and integrated systems. In this section, we will explore key concepts like **Pulse Width Modulation (PWM)**, **wireless communication**, and **integrating Arduino with other platforms** such as **Raspberry Pi**.

1. Pulse Width Modulation (PWM)

Pulse Width Modulation (PWM) is a powerful technique used to simulate analog output with a digital signal. It is widely used in various Arduino projects for controlling things like motor speed, LED brightness, and even audio signals. PWM works by adjusting the width of the pulse (on/off period) sent through a digital pin to control how much time the signal stays on versus off during each cycle.

How PWM Works:

PWM allows you to vary the average voltage supplied to a device without actually changing the supply voltage. For instance, when you use PWM to control the brightness of an LED, you're sending a high-frequency square wave to the LED. The longer the pulse stays "on" within a given cycle (the higher the duty cycle), the brighter the LED appears. Conversely, a shorter "on" time (lower duty cycle) makes the LED dimmer.

Key Applications of PWM:

- **Motor Speed Control:** By varying the pulse width, you can control the speed of motors in projects like robots or fans.
- **LED Brightness:** Use PWM to create smooth lighting effects by dimming LEDs.
- **Audio Generation:** Arduino can generate simple tones using PWM, often used in buzzers or small speakers.

How to Use PWM on Arduino:

On Arduino, PWM is available on most digital pins, and it's as simple as using the analogWrite() function to set a duty cycle. For example, to dim an LED connected to pin 9, you can use the following code:

2. Wireless Communication

Wireless communication is essential in modern electronics and allows your Arduino to communicate with other devices without the need for physical wires. There are several ways to achieve wireless communication with Arduino, including **Bluetooth**, **Wi-Fi**, **RF (Radio Frequency)**, and **Zigbee**.

Types of Wireless Communication:

- **Bluetooth (HC-05/HC-06 Modules):** Bluetooth allows for simple two-way communication with a smartphone or another Bluetooth-enabled device. This is ideal for building remote-control projects like robots or smart home devices.
- **Wi-Fi (ESP8266/ESP32):** The ESP8266 and ESP32 are powerful Wi-Fi modules that can connect Arduino to the internet. They enable your Arduino projects to send data to the cloud, control devices remotely over the internet, or even interface with web-based services.
- **RF Modules (nRF24L01, RF433):** RF communication is great for low-power, short-range communication. You can use RF modules to send and receive data between multiple Arduino boards in projects like remote sensors or wireless data logging.

Example of a Wi-Fi Project (ESP8266):

Suppose you want to control an LED remotely over the internet using the ESP8266 module. Here's how you could set it up using Arduino and the ESP8266.

1. Connect the **ESP8266** module to your Arduino using the appropriate pins.
2. Install the necessary libraries (ESP8266WiFi and WiFiClient).
3. Use the following code to connect to Wi-Fi and control an LED over the internet.

3. Integrating Arduino with Raspberry Pi

While Arduino is great for low-level hardware control, **Raspberry Pi** is a full-fledged computer with more processing power and capabilities like running Linux, handling multimedia, and supporting advanced networking. Integrating Arduino with Raspberry Pi allows you to leverage the best of both worlds: the simplicity and precision of Arduino for physical computing tasks and the computing power of Raspberry Pi for more complex processing, internet connectivity, and data management.

Why Integrate Arduino with Raspberry Pi?

- **Data Processing:** Offload heavy data processing to the Raspberry Pi, while the Arduino handles sensor readings and device control.
- **Advanced Communication:** Use the Raspberry Pi to handle internet-based tasks, like cloud computing, while Arduino handles local control.
- **Multiple Interfaces:** Use Raspberry Pi's HDMI, USB, or Ethernet ports to interact with users or connect to external systems.

Example of Arduino and Raspberry Pi Integration:

Let's say you want to build a weather station that logs data from sensors using Arduino but stores and displays the data on a web page hosted by the Raspberry Pi.

1. **Arduino:** Read sensor data (temperature, humidity, etc.) and send it to the Raspberry Pi via **serial communication**.
2. **Raspberry Pi:** Run a Python script that reads the serial data from the Arduino and logs it to a database or displays it on a webpage.

Provide Access to Resources: Tools for Advanced Learning

To truly master Arduino and take your skills to the next level, it's essential to continue learning from a variety of sources. The Arduino community is vast, and there are many online resources, tutorials, and forums where you can access advanced projects, solutions, and support.

Here are some valuable resources to help you along the way:

- **GitHub:** GitHub is home to countless open-source Arduino projects. You can find libraries, code examples, and even entire projects to help you learn and experiment. Search for repositories related to your interests, such as wireless communication, robotics, or IoT.
- **Arduino Forums (forum.arduino.cc):** The official Arduino forum is a great place to ask questions, share your projects, and collaborate with other Arduino

enthusiasts. It's a goldmine for troubleshooting tips and project ideas.
- **Adafruit Learning System:** Adafruit offers an incredible range of tutorials, from beginner to advanced. They cover topics like sensors, IoT, and advanced electronics.
- **Instructables:** Instructables is a platform where makers and creators share their DIY projects, many of which include Arduino-based designs. You can find step-by-step guides for advanced Arduino projects here.
- **Hackster.io:** Hackster.io is another community-driven platform where users post their projects. Many of these are open-source and come with detailed tutorials to help you build complex systems with Arduino and other hardware.
- **YouTube Channels:** Many Arduino experts and enthusiasts share their projects and tutorials on YouTube. Channels like **Paul McWhorter**, **Jeremy Blum**, and **GreatScott!** provide detailed explanations of advanced concepts.

By taking advantage of these resources, you can continually expand your knowledge and stay up-to-date with the latest advancements in Arduino technology.

By mastering advanced concepts like PWM, wireless communication, and integration with Raspberry Pi, you'll be well on your way to creating more powerful, sophisticated, and interactive Arduino projects. Remember, the key to success is continuous learning and hands-on experimentation—so don't hesitate to explore, collaborate, and push the boundaries of what's possible with Arduino.

Chapter 9: Troubleshooting Like a Pro: Diagnosing and Solving Common Issues

No matter how experienced you are, encountering problems in your Arduino projects is inevitable. Whether you're just starting out or working on advanced builds, troubleshooting is a vital skill. Learning how to diagnose and solve issues efficiently can save time and prevent frustration. In this chapter, we'll guide you through common problems, teach you how to approach debugging systematically, and provide tips for resolving issues like a pro.

1. Understanding the Troubleshooting Process

The key to becoming a proficient troubleshooter is developing a structured approach. Rather than getting frustrated when something doesn't work, take a step back and break down the problem logically. Here's how you can approach the troubleshooting process:

- **Identify the Problem:** The first step is to figure out what's going wrong. Does the Arduino not power on? Is your LED not lighting up? Are sensors failing to provide readings? Is the motor not spinning? The more specific you are about the problem, the easier it will be to diagnose.
- **Isolate the Issue:** Once you have a general sense of the problem, isolate the issue to a particular component or area of the code. Is it a hardware issue, a wiring problem, or a software bug? Is the problem with the Arduino board itself, the components (sensors, actuators), or the connections?

- **Check Your Code:** Before diving into hardware, always start with your code. It's easy to overlook small mistakes in programming that can cause issues, like incorrect syntax, incorrect logic, or forgotten libraries.
- **Inspect the Hardware:** If your code seems correct, it's time to check the hardware. Look for poor connections, incorrect wiring, or faulty components. Ensure that your breadboard, jumper wires, and Arduino pins are properly connected.
- **Test in Phases:** If your project is large, break it down into smaller, manageable sections. Test each part of the circuit and code independently to see where the failure occurs. By isolating sections, you'll be able to pinpoint exactly where the problem lies.

2. Common Issues and How to Solve Them

Let's go through some of the most common problems that Arduino users face and how to troubleshoot them.

a) Arduino Won't Power On

Problem: You upload code, but nothing happens—the Arduino board doesn't light up, and no response is visible.

Possible Causes:

- **Power Source Issue:** Ensure your Arduino is powered. If you're using a USB cable, check that it's properly connected to the computer. If you're using an external power source, make sure the voltage is correct.

- **Faulty Cable or USB Port:** Sometimes the issue is as simple as a damaged USB cable or faulty port. Try using a different cable or port.
- **Blown Fuse or Faulty Arduino Board:** In rare cases, the Arduino board itself may be damaged. Test it with a different cable or check with another Arduino board to ensure your hardware is functioning.

Solution:

- Try using a different USB cable or power supply.
- Check the onboard LED; if it's blinking or on, it indicates that power is reaching the board.
- If using external power, check the voltage with a multimeter to ensure it's within the recommended range.

b) Serial Monitor Isn't Displaying Output

Problem: You upload a program that prints to the serial monitor, but nothing appears on the screen.

Possible Causes:

- **Incorrect Baud Rate:** If the baud rate in your Serial.begin() function doesn't match the baud rate in your Serial Monitor, you won't see any data. The default baud rate is 9600, but make sure it's set consistently in both places.
- **Incorrect Serial Port:** Ensure that you've selected the correct COM port in the Arduino IDE. Sometimes the wrong port can be selected, and data won't appear in the monitor.

- **Code Issue:** If there's a bug or infinite loop in your code, the serial output may not be triggered.

Solution:

- Double-check the baud rate in your Serial.begin() and ensure it matches the setting in the Serial Monitor.
- Go to **Tools > Port** in the Arduino IDE and select the correct COM port.
- If there's a code issue, use delay() to slow down loops and allow output to be sent to the serial monitor.

c) LED Doesn't Light Up

Problem: You're working on a project where an LED should light up, but it's staying off.

Possible Causes:

- **Wrong Wiring:** The anode (positive leg) should be connected to the output pin, and the cathode (negative leg) to ground. If the connections are swapped, the LED won't light up.
- **Current Limiting Resistor:** If you don't use a current-limiting resistor (typically 220Ω to 1kΩ), the LED could be damaged, or the circuit may not function correctly.
- **Incorrect Pin Mode or Code Logic:** Make sure that the pin you've connected the LED to is configured as an OUTPUT in your code.

Solution:

- Check the wiring, ensuring the positive lead is connected to the Arduino pin and the negative lead is connected to ground.
- Use a 220Ω or 330Ω resistor in series with the LED to limit current and protect both the LED and the Arduino.
- Review the code to ensure the pin mode is set to OUTPUT and that you're sending a high signal (HIGH) to the LED pin.

d) Sensor Values Are Incorrect or Fluctuating

Problem: You connect a sensor (like a temperature or light sensor), but the readings are inconsistent or don't match expected values.

Possible Causes:

- **Improper Wiring:** Ensure that the sensor is wired properly—check the datasheet for pinout information.
- **Sensor Calibration:** Some sensors require calibration to provide accurate readings. For instance, a temperature sensor might need to be adjusted for accurate temperature measurement.
- **Noise or Interference:** Analog sensors (like temperature sensors) are often sensitive to electrical noise, which can cause fluctuating readings.

Solution:

- Double-check the wiring, especially if your sensor has multiple pins. Make sure it's connected to the right pins (e.g., ground, power, and signal).

- If using an analog sensor, consider adding a capacitor (typically 100nF) between the sensor's power and ground pins to smooth out noise.
- For temperature sensors, use known calibration values or compare the readings with an accurate thermometer to check for discrepancies.

e) Motors Aren't Running

Problem: You're using a motor (e.g., a DC motor, servo, or stepper), but it isn't responding to the code.

Possible Causes:

- **Insufficient Power:** Motors often require more current than the Arduino can supply through its onboard pins. Ensure the motor has its own power source if necessary.
- **Motor Driver Issues:** Many motors require a motor driver to work with Arduino. If you're using a motor driver (e.g., L298N for DC motors), ensure it is wired correctly and receiving the appropriate signals.
- **Code Logic or Pin Assignment:** Make sure the motor pins in your code match the ones where the motor driver is connected.

Solution:

- Use an external power supply to power the motor (check the motor's voltage and current ratings).
- If using a motor driver, consult the datasheet and wiring diagrams to make sure the connections are correct.

- Debug the code by simplifying it to a basic test to ensure the motor is being controlled properly.

3. Tools and Tips for Effective Troubleshooting

To troubleshoot efficiently, certain tools and techniques can be invaluable:

- **Multimeter:** A multimeter is one of the most powerful tools for debugging electronics. It can help you check voltages, resistances, and continuity. Use it to check the power supply, the status of pins, and the wiring between components.
- **Arduino Serial Monitor:** Always check the Serial Monitor for output and error messages. Print variables, sensor readings, or status messages to help you understand where the code is failing.
- **LED Indicators:** In more complex circuits, using additional LEDs to indicate different states (e.g., debugging status) can help identify where in the circuit the issue arises.
- **Testing in Phases:** If your project is large, break it down into smaller parts and test each section separately. For example, first test your sensors, then your motors, then the communication between components. This will help you isolate where things are going wrong.

4. Building Confidence with Troubleshooting

Every time you troubleshoot and solve a problem, you gain confidence and learn more about the inner workings of your Arduino projects. Mistakes and errors are a natural part of the learning process, and troubleshooting allows you to deepen your understanding of both hardware and software.

Remember, the most successful makers are the ones who aren't afraid to tackle problems head-on. When you run into an issue, take a methodical approach, and don't be afraid to ask for help from the Arduino community.

With patience, perseverance, and the right techniques, you'll quickly master the art of troubleshooting, making you a more skilled and confident Arduino maker.

In this chapter, we've covered some common issues and outlined how to diagnose and fix them efficiently. By following a systematic approach, leveraging the right tools, and building confidence in your problem-solving abilities, you'll soon be troubleshooting like a pro!

Real-World Problems: Common Issues Beginners Face and How to Solve Them

As a beginner in the world of Arduino programming and electronics, you'll inevitably encounter challenges. However, these challenges provide valuable learning experiences that will help you grow as a maker. In this section, we'll explore some of the most common issues beginners face, such as powering up the Arduino, wiring problems, and other hardware-related challenges. We'll also walk you through step-by-step solutions to help you troubleshoot and resolve these issues, ensuring you gain the confidence to tackle future projects.

1. Problem: Powering Up Your Arduino – No Response

Story:
Sarah just received her first Arduino starter kit and was eager to try out a simple project. She plugged the Arduino into her computer via the USB cable, opened the Arduino IDE, and attempted to upload a basic "Blink" program to make the onboard LED blink. However, nothing happened. The Arduino didn't power on, no lights flickered, and no response showed up on the Serial Monitor.

Step-by-Step Solution:

- **Check the USB Cable and Port:** One of the most common causes for the Arduino not powering up is a faulty USB cable or USB port. Sometimes, the cable might be for charging only, and not for data transfer. Try using a different USB cable that supports data transmission and test it on another port or computer to ensure the cable and port are functional.
- **Verify the Power Source:** If you're using an external power source, ensure that it's within the recommended voltage range (7-12V for most Arduino boards). If the power supply is incorrect or underpowered, the board won't turn on.
- **Look for an Indicator Light:** Most Arduino boards have a built-in LED that should light up when powered on (usually labeled "ON" or "PWR"). If this LED doesn't light up when you plug the board into your computer, there could be an issue with the board itself.
- **Check the Board and Connections:** Inspect the board for visible signs of damage such as burnt

components or loose connections. If the board is new and nothing happens after trying the above steps, there might be an issue with the board itself.

2. Problem: Incorrect Wiring – LED Doesn't Light Up

Story:
John followed a tutorial to create a simple project where an LED blinks on and off. After wiring everything up, including a 220Ω resistor to limit the current to the LED, he uploaded the code. However, the LED remained off, despite the code being correct and the Arduino powering on.

Step-by-Step Solution:

- **Check the LED Polarity:** LEDs are polarized, meaning they only work when connected in the correct direction. The longer leg of the LED is the positive (anode) leg, and it should be connected to the Arduino output pin. The shorter leg is the negative (cathode) leg, and it should go to ground. If the LED is reversed, it won't light up.
- **Ensure the Correct Resistor Value:** The resistor connected in series with the LED is essential to limit the current and prevent the LED from burning out. If John used a resistor that was too small, such as a 10Ω resistor, too much current would flow through the LED, potentially damaging it. The correct value should be 220Ω to 330Ω for most common LEDs.
- **Check Wiring Connections:** Sometimes, even when the components are correctly chosen, a loose or incorrect wire connection can cause problems. Ensure the wiring is solid and correctly placed, as a

poor connection between the Arduino pin and the LED will prevent it from lighting up.
- **Test the LED with Simple Code:** To eliminate the possibility of a programming issue, John can upload a basic sketch to just turn the LED on and off (without any delay or logic) to confirm whether it's a hardware issue. This isolates the problem to the wiring and not the code.

3. Problem: No Serial Monitor Output – Code Is Running but No Data Shows

Story:
Emily was following a project that displayed sensor readings on the Serial Monitor. She connected a temperature sensor to the Arduino, uploaded the code, and opened the Serial Monitor. However, she saw no data, even though the code was running.

Step-by-Step Solution:

- **Ensure the Correct Baud Rate:** The baud rate in the Serial.begin() function must match the baud rate in the Serial Monitor. If the baud rate is mismatched, the Serial Monitor won't display anything. Emily should ensure that the baud rate in the code (e.g., Serial.begin(9600);) matches the baud rate setting in the Serial Monitor (also set to 9600).
- **Select the Right COM Port:** Sometimes, the correct COM port isn't automatically selected in the Arduino IDE. Emily should go to **Tools > Port** and select the correct COM port associated with her Arduino.
- **Check the Code Logic:** If the baud rate and COM port are correct, the issue might be within the code

itself. Emily should check if the Serial.print() commands are inside the right loop or conditional statements. If there's a while loop or if statement that isn't being triggered, the Serial Monitor may remain empty.
- **Test with Basic Code:** Emily can simplify the code to a basic program that prints something simple like "Hello World" to the Serial Monitor to confirm that the communication between the Arduino and the Serial Monitor is working.

4. Problem: Circuit Doesn't Work – Sensor Not Giving Readings

Story:
Michael was excited to make a light-sensitive project using a photoresistor (LDR). He connected the sensor to the analog input pin, uploaded the code to read light intensity, and opened the Serial Monitor to see the results. However, all he got was zero values, even though the room light was changing.

Step-by-Step Solution:

- **Check Wiring Connections:** Michael should start by checking the wiring of the LDR. Typically, an LDR has two pins, one connected to the analog input pin of the Arduino and the other connected to ground. A pull-down resistor (typically 10kΩ) is used to ensure the correct voltage is read. Without this resistor, the input pin might float and produce incorrect values.
- **Verify the Sensor's Orientation:** LDRs are light-sensitive, so Michael needs to ensure that it's positioned correctly to detect changes in light. If the

LDR is facing a shadowed area or isn't aligned with the light source, it may return a constant zero reading.
- **Test with Known Values:** To verify if the sensor is working, Michael can check its output by using the analogRead() function to print out the sensor's raw values. If they fluctuate between 0 and 1023, the sensor is likely functioning properly. If the readings stay stuck at zero, the wiring or sensor could be faulty.
- **Test the Sensor with Different Light Conditions:** LDRs are sensitive to light, so testing with different light intensities (e.g., covering it with your hand, shining a flashlight on it) can help determine if the sensor is responding to light changes. Michael should try this and check the readings again.

5. Problem: Motor Won't Turn – DC Motor Not Responding

Story:
Lisa followed a tutorial to create a motorized project using a DC motor. She connected the motor to the Arduino, wrote the code to turn the motor on, and ran the program. However, the motor didn't move, despite everything being set up correctly.

Step-by-Step Solution:

- **Check the Motor Power Supply:** Many motors, particularly DC motors, require more power than an Arduino can provide through its pins. Lisa should check whether the motor is connected to an external power supply (e.g., a 9V battery or external adapter) instead of the Arduino's 5V pin.

- **Use a Motor Driver:** DC motors typically require a motor driver (e.g., L298N or L293D) to control the motor speed and direction. Without a motor driver, the Arduino may not be able to supply enough current to operate the motor. Lisa should confirm that she's using a motor driver and that it's wired correctly.
- **Check the Motor Connections:** If Lisa is using a motor driver, she should make sure the motor is connected to the correct terminals on the motor driver. Incorrect wiring of the motor to the driver could prevent it from spinning.
- **Verify Code and Pin Assignments:** Sometimes, the motor won't turn because the correct pins in the code aren't matched to the Arduino pins. Lisa should check the code to ensure the correct pins are being used to control the motor driver's inputs.

DIY Fixes: Diagnosing Code Issues, Broken Circuits, and Malfunctioning Components

When working with Arduino projects, it's common to run into problems—whether it's an issue with the code, a broken circuit, or malfunctioning components. However, the ability to troubleshoot and fix these issues is one of the most empowering skills you can develop as an Arduino enthusiast. In this section, we'll explore some practical DIY fixes to common problems you might face when working with Arduino, along with tips on how to diagnose these issues effectively. We'll also provide techniques to help you think through problems systematically, so you can solve them independently.

1. Diagnosing Code Issues: Why Doesn't My Program Work?

When things go wrong in your Arduino project, the first place to check is the code. An error in the code is often a subtle issue that might be tricky to spot, but with a systematic approach, you can fix it.

DIY Fixes:

- **Use the Serial Monitor:** The Arduino Serial Monitor is your primary debugging tool. If your program isn't behaving as expected, print out values at different stages of the code to see where things are going wrong. For example, if you're controlling a motor and it's not spinning, use Serial.print() to check if the right signals are being sent to the motor control pins.

   This will allow you to track the value of motorSpeed and determine if it's what you expect. If the value doesn't change, you know there's an issue earlier in the code.

- **Check for Syntax Errors:** Syntax errors are common in programming. These could be something as simple as a missing semicolon, an incorrect bracket, or a typo in a function name. The Arduino IDE typically highlights syntax errors in red, so be sure to carefully review the error messages.
- **Verify Pin Definitions:** Ensure that the pins you've assigned in the code match the physical connections of your components. This includes checking for any confusion between digital and analog pins. A mismatch in pin definitions could cause certain

components to remain inactive, even though the code seems fine.
- **Simplify Your Code:** If your code is complex, break it down into smaller parts. Test smaller chunks of the code one at a time. If something isn't working, isolate the problematic part by commenting out sections of code and running only the critical part. This approach can help you pinpoint where things are going wrong.
- **Libraries and Dependencies:** Many Arduino projects rely on external libraries for components like sensors, displays, and motors. If your code uses a library, make sure it's installed correctly. Sometimes an outdated or incompatible library can break the code, so check for any available updates.

2. Fixing Broken Circuits: When the Wiring Goes Wrong

Wiring problems are one of the most common sources of frustration for beginner Arduino users. A loose connection, a misplaced wire, or an incorrect jumper can cause your circuit to fail. Here's how you can troubleshoot wiring issues.

DIY Fixes:

- **Check for Loose Connections:** A loose wire or poor connection could be the culprit. It's worth double-checking each connection in your circuit, especially on a breadboard. Gently press on the jumper wires to make sure they are firmly seated. If a connection is weak or wobbly, it can result in erratic behavior or failure to function.
- **Use a Multimeter:** A multimeter is a simple but powerful tool for diagnosing circuit issues. Use it to

check if the power supply is reaching the components. For example, you can measure the voltage on a pin to confirm whether your Arduino is providing the correct signal. You can also use it to check for continuity, ensuring there are no breaks in your wires or connections.

How to Use a Multimeter:

1. Set the multimeter to measure voltage (DC).
2. Place the multimeter probes on the ground and the pin you're testing.
3. Check the reading; it should match the expected voltage.

- **Double-Check Component Placement:** When working with components like sensors, LEDs, or motors, ensure they are in the right orientation. For example, LEDs have polarity; the longer leg should go to the positive side of the circuit. Reversing components can prevent the circuit from working properly or even damage them.
- **Inspect the Power Source:** If your circuit isn't powering on, make sure that your power source is working. If you're using a battery, check the voltage with a multimeter to confirm it's not drained. If you're using an external power supply, verify that it's within the appropriate voltage range (e.g., 7–12V for most Arduino boards).
- **Look for Short Circuits:** Sometimes wires can accidentally touch each other, creating a short circuit. Look closely for any areas where wires are crossing or components might be touching each other. A short circuit can cause the board to behave erratically or even damage the components.

3. Fixing Malfunctioning Components: What If the Sensor, Motor, or LED Isn't Working?

Components, even high-quality ones, can sometimes malfunction. Whether it's a sensor that's not responding, an LED that won't light up, or a motor that's stuck, here's how to troubleshoot common component issues.

DIY Fixes:

- **Test Components Individually:** To determine if a component is faulty, test it in isolation. For example, test a new LED with a basic code to check if it lights up. If it doesn't, it might be defective. Similarly, test sensors by reading values from them directly via the Serial Monitor. If the sensor doesn't provide meaningful readings, it could be broken.
- **Swap Components:** If you suspect a component is malfunctioning, swap it with another working one. For example, replace the LED with a known good one, or try a different motor or sensor. This process helps you narrow down whether the problem lies in the component or somewhere else in the circuit.
- **Check Component Specifications:** Ensure that the component you're using is compatible with your Arduino board. For example, some sensors require specific voltage levels (e.g., 3.3V vs. 5V) and might not work correctly if you provide them with incorrect voltage. Always check the datasheet for each component before using it.
- **Inspect for Visible Damage:** Physical damage to components—such as burned-out chips, cracked sensors, or broken wires—can often be spotted. If you notice any component that looks damaged, it's likely

the cause of the issue, and replacing it with a new one should solve the problem.
- **Component Pins and Leads:** Some components, especially sensors and integrated circuits (ICs), have multiple pins that need to be connected to specific pins on the Arduino board. A mismatch in pin connections could cause the component to malfunction. Refer to the datasheet and verify the pinout before wiring up the component.

Empowering Problem-Solving: Building Confidence Through Troubleshooting

Learning how to diagnose and solve problems is a critical part of mastering Arduino. The more you troubleshoot, the more intuitive it becomes. Here are some tips to empower you to solve problems independently and boost your confidence in working with Arduino:

Focus on the Process:

- **Don't Panic, Stay Methodical:** When troubleshooting, it's easy to feel overwhelmed. Take a deep breath and tackle one problem at a time. Start by identifying the issue, and then break it down into smaller components (e.g., code, wiring, power supply, components). By staying systematic, you can prevent yourself from getting lost in the complexity of the issue.
- **Use the Scientific Method:** Think like a scientist—make a hypothesis, test it, and analyze the results. For example, if you suspect the motor is faulty, test it by swapping components and confirming if the motor

works in a different setup. By isolating variables, you can confidently identify the cause of the problem.
- **Document Your Process:** Keeping a troubleshooting log can be immensely helpful. Write down the steps you took, what you tested, and the results. This will not only help you keep track of your progress but also act as a reference in case you encounter the same problem in the future.
- **Learn from Mistakes:** Don't be afraid to make mistakes. In fact, it's an essential part of the learning process. Each failure is an opportunity to gain more knowledge and refine your problem-solving skills. As you encounter and fix problems, you'll become more resourceful and capable.

Chapter 10: Moving Forward: How to Keep Learning and Innovating with Arduino

As you finish your first few Arduino projects, you may find yourself with a sense of accomplishment, but also a growing curiosity about what's next. What new horizons can you explore with this remarkable platform? How can you continue to push the boundaries of your creativity, and how do you take your learning to the next level? This chapter is all about guiding you to continue your Arduino journey beyond the basics and keep innovating, learning, and exploring.

Whether you're transitioning from a beginner to an intermediate user or already thinking about advanced projects, the key to growth in the Arduino world lies in continued learning, community engagement, and creative problem-solving. In this chapter, we'll explore strategies for expanding your knowledge, staying inspired, and continuously pushing the limits of what you can achieve with Arduino.

1. Embrace Lifelong Learning: Constantly Evolving Your Skills

Arduino, and the larger world of electronics, programming, and robotics, is always evolving. New components, libraries, software updates, and project ideas emerge all the time. To keep growing and evolving, it's crucial to adopt a mindset of lifelong learning. The more you learn, the more you'll realize there's always something new to explore.

How to Keep Learning:

- **Follow New Trends and Technologies:** Keep an eye on the latest trends in Arduino and the maker community. New technologies, such as advanced sensors, AI integration, machine learning, and IoT (Internet of Things), are all becoming increasingly accessible to hobbyists and creators. By staying updated with these trends, you'll know what new possibilities are available for your projects.
    - **Example Resources to Explore:**
        - *Arduino Blog:* Stay informed about new projects, tutorials, and hardware releases.
        - *Maker Faires:* These events are great for seeing cutting-edge projects and networking with like-minded individuals.
        - *Online Tech News:* Websites like Hackaday and Instructables feature new Arduino projects, tutorials, and breakthroughs every day.
- **Take Online Courses:** There are countless online platforms that offer free and paid Arduino courses to help you move from beginner to expert. Websites like Udemy, Coursera, edX, and even Arduino's official website feature comprehensive courses on various Arduino-related topics. By completing these courses, you'll be exposed to new techniques and tools to add to your skill set.
- **Books and eBooks:** Never underestimate the power of a well-structured, detailed book. Books provide deep dives into specific topics that you might not easily find through online tutorials. Books on topics like advanced robotics, machine learning, and IoT

often feature project examples that push the boundaries of what you can build.
- **Experimentation and Play:** At the core of learning with Arduino is experimentation. Don't just follow tutorials—try modifying projects, changing sensors, or adding your own twist to them. Experimentation will not only help you discover new techniques but will also fuel your creativity. Break things, troubleshoot them, fix them, and see what happens. The lessons you learn in the process are irreplaceable.

2. Build a Portfolio of Projects: Show Off Your Skills

One of the best ways to solidify your learning and track your growth is by creating a portfolio of your projects. Whether it's a simple temperature sensor, a smart home automation project, or a fully functional robot, every project you complete adds to your experience and demonstrates your growing skill set. A portfolio helps you not only reflect on how much you've learned but also serves as a great showcase for potential employers, collaborators, or clients.

How to Build a Portfolio:

- **Document Your Projects:** Whenever you complete a project, make sure to document it in detail. Include:
    - A project overview (what it does and how it works)
    - The hardware used (Arduino board, sensors, motors, etc.)
    - The software you've written (with explanations of the key code elements)
    - Challenges faced and how you solved them

- Photos, videos, and even circuit diagrams to illustrate your work
- **Create a Personal Website or Blog:** Use platforms like WordPress, Wix, or GitHub Pages to create a website or blog where you can showcase your Arduino projects. This not only serves as a portfolio but can also help you connect with others in the maker community. You can use your website to write about the process, challenges, and solutions for each project, as well as share code snippets or tutorials.
- **GitHub:** For sharing code and collaborating with others, GitHub is invaluable. It's an online platform where you can host and manage your code, share your projects with others, and contribute to open-source projects. By contributing to Arduino projects or releasing your own code, you'll build credibility in the community while keeping a record of your learning journey.
- **Social Media:** Platforms like Instagram, Twitter, and TikTok are also great places to share your projects. You can post photos and short videos demonstrating your work, and even reach a wide audience interested in DIY electronics. Many makers use these platforms to document their progress and share their successes and setbacks.

3. Engage with the Arduino Community: Collaborate and Share Ideas

The Arduino community is vibrant, diverse, and full of innovative people with a wide range of skills and knowledge. Engaging with this community is one of the best ways to keep learning and growing, as you can share ideas, get feedback, and be inspired by others.

How to Engage:

- **Join Online Forums and Groups:**
    - *Arduino Forum*: The official Arduino forum is a great place to ask questions, find answers, and share your projects. You can seek help on technical issues or browse through existing threads to learn from others' experiences.
    - *Reddit*: Subreddits like r/arduino and r/maker are popular places to share your projects, get feedback, and ask for advice.
    - *Discord Servers*: Many Arduino and electronics-focused Discord communities exist where you can interact with other makers in real-time.
- **Participate in Maker Faires and Competitions:** Maker Faires, hackathons, and DIY electronics competitions are great ways to meet other enthusiasts, learn new skills, and showcase your projects. These events are full of inspiration and can motivate you to push the boundaries of your own projects.
- **Collaborate on Open-Source Projects:** Contributing to open-source Arduino projects is a great way to learn advanced techniques and collaborate with others. Websites like GitHub feature thousands of Arduino-related open-source projects, and contributing to these projects will expose you to new tools, techniques, and technologies.
- **Follow Arduino Influencers and Experts:** Many experts and experienced hobbyists share their knowledge on YouTube, blogs, and social media. Following these influencers can help you stay

updated on the latest trends, tools, and project ideas in the Arduino ecosystem.

4. Take on Bigger, More Challenging Projects: Push the Boundaries

Now that you've mastered the basics, it's time to take on more ambitious, challenging projects. The key to continuous growth is stepping outside of your comfort zone and tackling more complex tasks.

How to Take Your Projects to the Next Level:

- **Explore Robotics:** Now that you're comfortable with sensors, motors, and coding, why not design and build robots? From basic line-following robots to more advanced autonomous machines, robotics combines a variety of skills including coding, electronics, and mechanical engineering.
- **Build Internet of Things (IoT) Projects:** With IoT, you can create connected devices that communicate over the internet. Projects like a weather station that reports data to the cloud, a smart plant-watering system, or a home automation system can push your skills to the next level.
- **Learn Wireless Communication:** Experimenting with wireless technologies like Bluetooth, Wi-Fi, and RF (Radio Frequency) allows you to build more sophisticated, remotely controlled projects. Think about creating a Bluetooth-controlled robot, or a home security system with remote monitoring.
- **Work with Other Platforms:** Integrating Arduino with platforms like Raspberry Pi or microcontrollers like ESP32 opens up a world of possibilities. Arduino can

handle physical inputs/outputs while Raspberry Pi can handle more complex processing, networking, and communication tasks. This combination allows you to create even more powerful projects.
- **Explore Artificial Intelligence and Machine Learning:** Advanced Arduino users are now experimenting with integrating machine learning algorithms. For example, using sensors to collect data and then using AI models to make decisions based on that data. While this is an advanced area, there are many resources and tutorials that can help guide you.

5. Stay Inspired: Keep the Creative Spark Alive

Maintaining motivation and inspiration is key to a long-lasting Arduino journey. Sometimes, the hardest part is staying engaged after the initial excitement fades. Here are a few ways to stay inspired and keep the creative ideas flowing:

- **Set Personal Challenges:** Give yourself goals or challenges to work towards. It could be something as simple as "I want to build a smart mirror," or "I'll create a weather station that tracks humidity and temperature." Challenges help direct your efforts and keep you motivated.
- **Create with a Purpose:** Try to think about how Arduino can solve real-world problems, both big and small. By designing projects that can help others, you'll stay motivated. Perhaps you could create a device to assist in home health monitoring or build a system that helps automate energy consumption.
- **Reflect on Your Journey:** Periodically look back at the projects you've completed and how far you've come. The progress you've made will help fuel your

drive to tackle even more complex and exciting projects.

Focus on Lifelong Learning: Continuing Your Arduino Journey with Advanced Tools

Arduino is just the beginning of a vast world of possibilities. As you become more comfortable with the basics of coding and hardware, it's essential to look forward and keep pushing your limits. Lifelong learning is the cornerstone of innovation, and in the world of Arduino, there are always new tools, techniques, and ideas to explore. In this section, we'll dive into how you can continue your journey beyond beginner-level projects, introducing you to advanced tools, sophisticated sensors, and new coding techniques that will help you push your creativity and technical skills to new heights.

1. Embrace Advanced Simulation Software: Testing Without Limits

As your projects become more complex, one of the most valuable tools you can integrate into your workflow is **simulation software**. Simulation tools allow you to test your Arduino circuits, code, and designs without needing physical components. This is especially helpful for understanding how systems will behave in real-world conditions, troubleshooting, and optimizing your designs before building them.

Benefits of Using Simulation Software:

- **Test Complex Circuits:** Simulators can emulate complex circuits that might be hard to build physically,

allowing you to experiment without worrying about physical limitations.
- **Save Time and Money:** Before purchasing components or assembling hardware, you can test your ideas in the software, which helps you avoid costly mistakes.
- **Learn Faster:** Simulators often provide a virtual environment where you can instantly see the effects of changes to your code or components, accelerating your learning process.

Popular Arduino Simulation Tools:

- **Tinkercad Circuits:** A great beginner-friendly platform that lets you simulate Arduino circuits and write code directly in the interface. It's perfect for testing out new projects quickly and getting hands-on experience with electronic components.
- **Proteus:** A more advanced simulator that allows for detailed circuit simulations and the ability to emulate both hardware and code. It's a great tool for complex projects that require high accuracy and reliability in simulations.
- **Fritzing:** Although more focused on prototyping, Fritzing also includes a simulation component that helps visualize your circuit and see how the Arduino board will interact with various components.

As you gain confidence, start integrating simulation tools into your workflow. They can help you test larger, more intricate systems and facilitate the trial-and-error process without the need for extensive physical setups.

2. Advanced Sensors: Enhancing Your Projects

One of the most exciting aspects of Arduino is the sheer variety of sensors available. As your projects grow more sophisticated, incorporating **advanced sensors** can open up new realms of possibility. These sensors allow your projects to interact with the physical world in deeper and more meaningful ways. By adding advanced sensors, you can gather data, track movement, sense the environment, and much more.

Types of Advanced Sensors to Explore:

- **Environmental Sensors:**
    - **Gas Sensors:** Used for detecting gases like methane or carbon monoxide. Useful for projects such as air quality monitoring or gas leak detection.
    - **Soil Moisture Sensors:** Perfect for creating automated irrigation systems for gardens or smart farming projects.
    - **UV Sensors:** Measure ultraviolet light, which could be used in solar-powered projects or UV exposure monitoring systems.
- **Motion and Proximity Sensors:**
    - **PIR Sensors (Passive Infrared):** Used to detect motion, which is perfect for building motion-activated lights or security systems.
    - **Ultrasonic Sensors:** Measure the distance between the sensor and an object, enabling projects like distance measurement tools or obstacle avoidance for robots.
- **Wearable Sensors:**

- **Heart Rate Sensors:** Can be integrated into wearable devices, creating projects like health-monitoring systems or fitness trackers.
- **Accelerometers and Gyroscopes:** Used to detect orientation, movement, and tilt. These sensors are essential for creating motion-based or robotic projects.

By integrating these advanced sensors into your projects, you'll be able to create smart, data-driven systems that react to the environment in real-time. This opens up vast opportunities for creating projects related to automation, smart homes, healthcare, robotics, and environmental monitoring.

3. New Coding Techniques: Elevating Your Programming Skills

As your hardware skills evolve, it's essential to upgrade your **coding techniques** to match. Arduino programming relies on a simplified version of C/C++, but as you dive into more advanced projects, you'll need to understand more intricate coding concepts to manage the complexity of your systems effectively.

Coding Concepts to Explore:

- **Object-Oriented Programming (OOP):** This programming paradigm is used to create reusable code by organizing it into objects (like classes). OOP can make your Arduino projects cleaner, more modular, and easier to maintain. By mastering concepts such as inheritance, polymorphism, and

encapsulation, you'll be able to write more organized and scalable code.
- **Example:** You could create classes to represent different sensors in your project (e.g., TemperatureSensor, MotionSensor), which would allow you to reuse code for similar sensors and manage multiple sensor inputs more efficiently.
- **Interrupts:** Interrupts allow you to temporarily pause the current execution of code to address a time-sensitive task. Learning to use interrupts can significantly improve the responsiveness of your projects, especially for real-time applications like monitoring or controlling devices with tight timing constraints.
  - **Example:** You can use interrupts to trigger actions in response to a button press or a sensor reading without having to constantly poll the sensor in a loop, improving efficiency.
- **State Machines:** A state machine is a way of managing complex systems by defining a set of states and the conditions for moving between them. State machines help you control devices with multiple modes or stages, such as a robotic arm that can move to different positions, or a home automation system with multiple modes of operation.
  - **Example:** A simple state machine could control an LED light that transitions through different colors based on sensor inputs or a timer.
- **Libraries and Frameworks:** As projects become more advanced, it's helpful to use libraries and frameworks to speed up development. There are a wide variety of open-source libraries for Arduino,

covering everything from sensor management to communication protocols like Bluetooth and Wi-Fi.
  - **Example:** The **Adafruit Sensor Library** simplifies interaction with various sensors, or the **WiFi101 Library** helps you quickly connect your Arduino to the internet.
- **Communication Protocols:** With the rise of IoT (Internet of Things) projects, understanding communication protocols such as **I2C**, **SPI**, **UART**, and **MQTT** will allow you to interface multiple components and build connected systems. Mastering these protocols is essential for projects involving multiple devices or remote communication.
  - **Example:** In a home automation system, you might use **I2C** to connect multiple sensors to an Arduino, while using **Wi-Fi** to control the system remotely via a mobile app.

4. Explore Integration with Other Platforms

While Arduino is powerful on its own, combining it with other platforms and microcontrollers can vastly expand your project's capabilities. **Integrating Arduino with other platforms** like **Raspberry Pi**, **ESP32**, and **FPGA** (Field-Programmable Gate Arrays) allows you to tackle even more complex projects by leveraging the unique strengths of each system.

How to Integrate Arduino with Other Platforms:

- **Raspberry Pi:** Raspberry Pi is a full-fledged computer with a Linux operating system, and it's great for projects requiring high computational power, networking, or multimedia. Arduino handles the

sensor inputs and outputs, while Raspberry Pi can be used for processing and communication. Combining these two systems can create powerful, connected devices.
  - **Example Project:** A home security system where Arduino controls the sensors (e.g., PIR for motion detection) and Raspberry Pi processes the camera feed and sends notifications.
- **ESP32:** An advanced microcontroller with Wi-Fi and Bluetooth capabilities, ESP32 can complement Arduino by providing wireless connectivity for IoT projects. It can handle more complex networking tasks while Arduino focuses on the hardware interface.
  - **Example Project:** A weather station where Arduino collects data from sensors, and ESP32 sends this data to the cloud for real-time monitoring.
- **FPGA:** If you are interested in hardware-level design and more complex processing tasks, learning FPGA programming can open up new possibilities. FPGAs are highly parallel, enabling you to create custom hardware solutions that outperform traditional microcontrollers in certain applications.
  - **Example Project:** Use FPGA for real-time image processing or complex signal manipulation, with Arduino managing peripheral devices and sensors.

5. Joining Communities and Networking: Expand Your Knowledge

As you dive into more advanced Arduino topics, it's essential to **connect with other like-minded individuals**. Joining online forums, participating in open-source projects, attending maker events, and contributing to GitHub repositories are all excellent ways to expand your knowledge.

Communities to Explore:

- **Arduino Forum:** A great place to ask technical questions and engage with both beginners and experts.
- **Reddit (r/arduino):** Share projects, ask questions, and get feedback.
- **Hackaday.io:** A platform for building and sharing open-source hardware projects.
- **Meetups and Maker Faires:** Local events where you can network with other makers, see new projects, and get inspired.

Personal Project Planning: Turning Your Arduino Ideas into Reality

As you continue exploring Arduino and its capabilities, you might feel inspired to create your own projects or even take your passion to the next level by starting a small Arduino-based business or community group. The power of Arduino lies not only in the ability to build cool gadgets but also in its potential to fuel entrepreneurial ventures, personal growth, and collaborative learning. Planning and executing a successful Arduino project requires careful thought, structure, and persistence. Here's a step-by-step roadmap to help you navigate the process from concept to creation, and even beyond to turning your hobby into something larger.

1. Define Your Idea: Identify Your Passion and Purpose

The first step in planning any Arduino project is to identify what excites you. What problem are you trying to solve? What challenge are you eager to tackle? Whether it's building a smart home device, creating a robotics project, or exploring wearable technology, start by narrowing down your project's scope.

Key Questions to Consider:

- **What problem are you solving?** Focus on real-world applications to ensure your project has a purpose. For instance, a home automation system could save energy or increase convenience.
- **What skills do you want to develop?** If you're interested in learning about sensors, opt for a project like a weather station or security system that uses a variety of sensors.
- **Who is your target audience?** Consider the end-users of your project. Is it for personal use, a gift, or a product you want to sell?

Project Idea Examples:

- **Smart Garden System:** Automate watering, monitor soil moisture, and control lighting, all powered by Arduino.

- **Health Monitoring Wearable:** Build a bracelet that monitors heart rate and sends data to your phone.
- **Bluetooth-Controlled Robot:** Design a mobile robot that you can control from your smartphone using Bluetooth.

Once you have a clear idea, outline the project's goals. What's the desired outcome? This helps in establishing a clear path toward completion.

2. Research and Gather Components

Once you know what you want to build, begin researching the necessary components. Whether you're creating a smart home device, a robotic system, or a sensor-based project, each project will have different hardware and software requirements.

Components Checklist:

- **Microcontroller/Board:** Arduino Uno, Nano, or Mega depending on your project size.
- **Sensors/Actuators:** Temperature, motion, light sensors, or motors for movement.
- **Communication Modules:** Bluetooth, Wi-Fi, or RF modules if you're incorporating wireless communication.
- **Power Supply:** Batteries, USB cables, or power adapters for long-term use.
- **Prototyping Materials:** Breadboards, jumper wires, resistors, capacitors, and other basic electronics.

Use online resources like the **Arduino Store**, **Adafruit**, or **SparkFun** to research and purchase the necessary components. Many of these stores also offer kits designed for specific projects, making it easier to get started.

3. Design the Project Architecture

Now it's time to create a **schematic**—a blueprint of your project. This should include the connections between various components such as sensors, motors, and the microcontroller. Tools like **Fritzing** can help you design a virtual circuit diagram. Additionally, creating a block diagram will allow you to visualize the flow of data or actions.

Consider These Elements:

- **Inputs and Outputs:** Identify which components will serve as inputs (e.g., sensors, buttons) and outputs (e.g., LEDs, motors).
- **Power Flow:** Plan how your system will be powered. Will it use batteries, solar power, or a wall adapter?
- **Communication:** Decide if your project needs to communicate wirelessly or through a wired connection, and choose the appropriate communication module.

This stage is all about planning and ensuring everything will fit together logically. A well-thought-out design helps you avoid unnecessary changes later in the project.

4. Write the Code: Develop and Test Iteratively

Start by writing your Arduino code in **small, manageable chunks**. Focus on one functionality at a time: first, test how your sensor reads data, then work on controlling an actuator, and finally, integrate everything into one cohesive system.

Keep your code modular so that you can troubleshoot and adjust easily.

Key Strategies for Writing Code:

- **Start Simple:** Break down your project into basic functions. Test each component independently before integrating them.
- **Use Libraries:** Arduino has a wealth of open-source libraries that help simplify coding. For instance, libraries for sensors, motors, and communication modules can save a lot of time.
- **Iterative Testing:** Test your code regularly during development. If your sensor doesn't work, debug and modify the code step by step. This approach will ensure that you catch issues early and learn through problem-solving.

Consider using **GitHub** to store your code and track changes. This can be a great way to collaborate with others and access a vast library of projects to learn from.

5. Build and Prototype: Start Assembling Your Components

Once your code is ready and your schematic is set, it's time to **assemble your hardware**. This is where prototyping materials such as breadboards, jumper wires, and breadboard-friendly components come into play. Start with a prototype before committing to a final design.

Key Tips:

- **Prototyping:** Begin by connecting your components on a breadboard. This makes it easy to swap out components or change the configuration without soldering.
- **Debugging:** As you assemble the circuit, make sure everything is connected correctly. Power up your system in stages—check sensor readings, motor actions, and communication before fully integrating.
- **Physical Design:** If your project requires a casing or enclosure, consider using **3D printing** or **laser-cut acrylic** to create custom housings. Design it to be functional yet visually appealing.

6. Test and Troubleshoot: Refine Your System

Testing is crucial for ensuring that your project functions as expected. When things don't work as planned, troubleshoot the hardware connections, review the code, and verify your power supply. Remember, debugging is part of the learning process and can often be the most rewarding part of creating a project.

Troubleshooting Tips:

- **Verify Power:** Ensure your power source is stable and providing the correct voltage to your Arduino and components.
- **Check Connections:** Double-check your wiring and ensure that each component is correctly connected according to the schematic.

- **Serial Monitor:** Use the **Arduino Serial Monitor** to log sensor data, debug output, and track the system's performance.

7. Going Beyond: Starting a Small Arduino-Based Business or Community

Once you've completed your personal project, you may want to share your work or turn it into something larger. Here are some ways you can take your Arduino experience beyond personal projects:

Start a Small Business:

- **Product Development:** If you've created a unique, functional product (e.g., a smart home device or an educational robot), consider prototyping and selling it. Start small with limited runs, and leverage platforms like **Etsy**, **eBay**, or **Kickstarter** to launch your product.
- **Consulting and Services:** Offer Arduino project design or consulting services to businesses or hobbyists. With enough expertise, you can help others develop their own projects or teach workshops.

Create a Community Group:

- **Maker Meetups:** Host or participate in **maker meetups** to showcase your projects, share knowledge, and collaborate with others.
- **Workshops:** Offer local Arduino workshops or online tutorials, helping others learn how to build their own projects. You could even create a YouTube channel to share your expertise and expand your audience.

- **Open-Source Projects:** Start an open-source initiative where you can collaborate with others, improve existing projects, or create new ones. Platforms like **GitHub** provide an easy way to share code and get feedback from the community.

Invitation to Stay Engaged: Connect with the Arduino Community

Arduino is more than just a tool—it's part of a **global community** of makers, tinkerers, and innovators. To continue growing as an Arduino enthusiast, it's essential to stay connected, learn from others, and share your own work. Here are some ways you can stay engaged:

Join Open-Source Platforms:

- **GitHub:** A vast platform where developers share their Arduino projects, libraries, and code. You can contribute to existing projects or start your own.
- **Hackaday.io:** A community platform for sharing hardware projects, from DIY electronics to complex robotics. It's a great place to learn from others and showcase your own projects.

Participate in Hackathons:

Hackathons are a fun and challenging way to push your Arduino skills further. Whether virtual or in-person, hackathons provide the opportunity to collaborate with other makers, solve real-world problems, and even win prizes. Websites like **Devpost** and **Hackathon.io** list events that are perfect for Arduino enthusiasts.

Engage with Online Social Groups:

- **Reddit (r/arduino):** A community where you can share your projects, ask for help, and discuss new developments in the world of Arduino.
- **Arduino Forum:** Join the official Arduino forum to engage with experts, troubleshoot problems, and discuss new project ideas.
- **Facebook Groups and Discord Servers:** Many groups dedicated to Arduino offer a space for collaborative learning, project sharing, and even networking with potential collaborators.

Never Stop Building

Whether you plan to continue creating personal projects, start a business, or lead a community, the opportunities for growth with Arduino are endless. Use the roadmap above to guide you in turning your ideas into reality and pushing the boundaries of what

Chapter 11: Bonus Chapter – Arduino and the Future: How This Technology is Shaping Tomorrow

In this bonus chapter, we step beyond the immediate applications of Arduino and look toward the future of this powerful technology. Arduino has already played a significant role in the DIY, educational, and prototyping worlds. However, the potential of this versatile platform is still expanding as new advancements in hardware, software, and the Internet of Things (IoT) create new opportunities for innovation. In this chapter, we'll explore how Arduino is influencing industries, education, and personal projects today and how it is shaping the future of technology.

1. Arduino and the Rise of the Internet of Things (IoT)

The Internet of Things (IoT) refers to the growing network of physical devices, vehicles, appliances, and other objects embedded with sensors, software, and connectivity that allows them to collect and exchange data. Arduino, with its flexibility and wide range of connectivity options, is at the heart of many IoT developments.

Arduino and IoT Integration

Arduino's open-source ecosystem is particularly well-suited for IoT projects, allowing developers and hobbyists to create connected devices without the need for complex systems. The platform provides a variety of modules and shields for wireless communication, such as Wi-Fi, Bluetooth, and LoRa (long-range radio communication). These features enable

Arduino to interface with the cloud, mobile devices, and other IoT systems.

Example Project:

- **Smart Home Automation**: Using Arduino, you can create a home automation system where lights, appliances, and even security cameras are controlled remotely via the internet. By integrating Arduino with cloud platforms like **ThingSpeak** or **Blynk**, users can manage their smart homes from anywhere in the world.

The Future of Arduino and IoT:
As IoT continues to expand, Arduino will play a central role in the development of affordable, flexible, and customizable IoT solutions. Industries ranging from agriculture to healthcare are leveraging Arduino to create smart systems that monitor conditions, provide data analytics, and automate processes in real time. The future will likely see even smaller, more powerful, and energy-efficient Arduino boards that can power a new generation of interconnected devices.

2. Arduino in Education: Democratizing Technology for Everyone

One of the most impactful ways Arduino is shaping the future is through its role in education. By providing a simple and cost-effective platform for learning electronics, coding, and problem-solving, Arduino has become an invaluable tool for teachers and students alike.

Arduino in STEM Education

Arduino allows students of all ages to explore concepts in **Science, Technology, Engineering, and Mathematics (STEM)** through hands-on projects. Its ease of use and wide range of tutorials make it accessible to beginners, while its flexibility and integration with advanced tools like sensors, actuators, and external modules appeal to experienced learners.

In schools, Arduino has been used to:

- Teach the basics of electronics, such as **circuit building** and **sensor integration**.
- Introduce students to **programming** with the Arduino IDE (Integrated Development Environment).
- Develop collaborative projects, encouraging teamwork and creative problem-solving.

Example Project in Education:

- **Interactive Learning Tools**: Arduino-based projects can help students explore physics concepts by building projects like a **light sensor** or a **temperature monitoring system**. These practical projects give students real-time data collection and analysis experience.

The Future of Arduino in Education:
In the coming years, we can expect more schools and universities to integrate Arduino-based courses into their curricula, including specialized programs for robotics, automation, and AI (Artificial Intelligence). With the rise of **virtual learning environments** and **maker spaces**,

students can even work on Arduino projects remotely, collaborating with peers from across the globe.

Additionally, Arduino's popularity in educational settings is helping to inspire a new generation of inventors, engineers, and problem solvers. In a world where technical skills are increasingly in demand, Arduino empowers students to develop the skills they need to succeed in tech-driven industries.

3. Arduino in the World of Robotics: A Path to Autonomous Systems

Robotics has been an area where Arduino's influence has truly blossomed. The platform's combination of accessibility, affordability, and extensibility makes it ideal for both hobbyist-level robots and prototypes for research and industry.

Arduino and Robotics Today

Arduino is widely used in building **autonomous robots**, **drone systems**, and **robotic arms**. Its capability to interface with various sensors, motors, and actuators allows developers to create robots that can interact with their environment, follow commands, and even perform complex tasks like object detection and navigation.

Example Project in Robotics:

- **Line Following Robot**: A simple yet effective example of Arduino's capabilities in robotics is the **line-following robot**, which uses infrared sensors to detect and follow a line on the ground. This project helps learners understand the principles of feedback loops, sensor integration, and basic movement algorithms.

The Future of Arduino and Robotics:
Arduino will continue to be a vital tool in advancing robotics technology. As **artificial intelligence** and **machine learning** become more integrated with hardware systems, Arduino boards will likely become more powerful and capable of handling sophisticated algorithms for robot decision-making. From **autonomous delivery drones** to **smart industrial robots**, Arduino-based systems will help shape the next generation of intelligent machines.

In addition, the low cost of Arduino boards and components opens up the possibility of mass participation in robotics development. This will lead to greater collaboration and innovation, as people from various backgrounds (engineering, art, medicine, etc.) contribute to the field of robotics.

4. Arduino and Sustainability: Creating Eco-Friendly Solutions

Sustainability is a growing concern across the globe, and Arduino can be a powerful tool for addressing environmental issues. With its ability to integrate sensors for monitoring environmental conditions and its capacity to control energy-efficient systems, Arduino is well-positioned to contribute to a more sustainable future.

Arduino and Environmental Monitoring

Arduino can be used in projects that help monitor and reduce the impact of human activity on the environment. For example, it can be used to create systems that:

- **Track air quality**, using sensors that detect pollutants and particulate matter.
- **Monitor water quality**, by measuring parameters such as pH levels and water temperature.
- **Optimize energy use** in homes or offices by controlling lighting and heating based on real-time conditions.

Example Project in Sustainability:

- **Solar-Powered Irrigation System**: Arduino can be used to design a solar-powered system that automatically waters plants based on soil moisture levels. This helps conserve water and ensures plants receive the necessary care without wasting resources.

The Future of Arduino and Sustainability:
As the world faces challenges like **climate change**, **resource depletion**, and **pollution**, Arduino will continue to be an invaluable tool for creating environmentally conscious technologies. The platform will enable more people to participate in green innovation, developing affordable solutions that help reduce waste, save energy, and monitor the state of the environment.

5. Arduino and the Future of Personal Innovation: Becoming a Creator of Tomorrow

While Arduino has already transformed the way hobbyists, educators, and professionals approach technology, it is also paving the way for a future in which **personal innovation** is not limited by large institutions or high costs. Arduino democratizes technology, making it accessible for anyone with a passion for building.

Personal Innovation and Entrepreneurial Opportunities

As more people embrace Arduino, we can expect to see an increase in **personal innovation**. Whether you want to develop a unique product, solve a specific problem, or build a business around a creative idea, Arduino provides the tools to bring your vision to life.

Example Project for Personal Innovation:

- **Wearable Tech**: Arduino allows creators to design wearable devices such as fitness trackers, sleep monitors, and even custom health devices. By combining sensors, Bluetooth modules, and small Arduino boards, individuals can build innovative products that meet specific personal needs or address niche markets.

The Future of Personal Innovation with Arduino:

As the cost of technology continues to drop and access to tools like 3D printing and Arduino development boards becomes more widespread, we'll see a rise in **independent inventors**. This could lead to new technologies that were once only dreamed of in science fiction, and individuals will play an integral role in shaping the future of technology.

Arduino is already a launchpad for many entrepreneurial ventures. As the platform grows and evolves, it will support **small businesses**, **startups**, and **hobbyists** to experiment with new ideas, prototype rapidly, and push the boundaries of what's possible.

Future Trends: The Future of Arduino in the Context of AI, Robotics, and Smart Cities

As technology continues to evolve at a rapid pace, Arduino's role is expanding far beyond basic projects. The open-source hardware and software platform is now deeply intertwined with cutting-edge fields like **artificial intelligence (AI)**, **robotics**, and **smart cities**. In this section, we'll explore how Arduino is positioned to drive innovation in these emerging technological trends and what the future holds for makers, hobbyists, and developers alike.

1. Arduino and Artificial Intelligence (AI): Empowering Intelligent Systems

Artificial Intelligence is one of the most transformative technologies of the 21st century, powering everything from **autonomous vehicles** to **smart assistants**. Arduino, with its ability to interface with sensors, motors, and other peripherals, has already begun to integrate AI concepts into DIY projects. While AI typically requires advanced computational power, Arduino is bridging the gap by enabling **edge AI** — processing and decision-making that occurs directly on the device rather than relying on cloud servers.

How Arduino is Enabling AI at the Edge

Arduino's processing power, though modest compared to traditional AI systems, is more than sufficient for many applications of AI, particularly in the **Internet of Things (IoT)** and **robotics**. Through platforms like **TensorFlow Lite**, Arduino boards can execute machine learning models directly on small, low-power devices. This is known as **edge computing**, which is becoming increasingly important as AI systems grow more decentralized and need to operate in real-time without latency caused by cloud processing.

Example Project:

- **AI-Powered Object Detection**: By integrating a **camera module** with a small Arduino-compatible board (e.g., **Arduino Portenta H7**), developers can create simple AI-powered systems for object recognition. These systems could be used in **robotics** (like autonomous navigation) or **smart security systems** that can identify people, animals, or objects in real time.

The Future of Arduino and AI:

In the future, Arduino will play an essential role in AI-driven projects by enabling:

- **Smart Sensors**: With AI models running directly on Arduino boards, sensors can make real-time decisions based on input data. This is useful for applications like **predictive maintenance**, where devices can predict failure based on patterns detected by sensors.

- **AI in Robotics**: Arduino-based robots could become more autonomous, using AI for tasks such as path planning, speech recognition, and even facial recognition for human-robot interaction.
- **Low-Cost AI Solutions**: Arduino's affordability allows small businesses and individual developers to experiment with AI applications without requiring powerful (and expensive) hardware.

As AI continues to evolve, Arduino will serve as an essential gateway to democratizing AI, providing creators with the tools to develop intelligent systems on a budget.

2. Arduino and Robotics: The Heart of Personal Robotics and Industry 4.0

Robotics is another rapidly growing field that Arduino is influencing. From **personal projects** to **industrial applications**, Arduino's ability to interface with motors, sensors, and actuators makes it an ideal platform for building robots. While many industrial robots are large and expensive, Arduino allows hobbyists, engineers, and researchers to develop sophisticated robots at a fraction of the cost.

Arduino in Robotics Today:

Currently, Arduino is used in a wide range of robotics applications, from **simple line-following robots** to **advanced autonomous systems**. By using a variety of sensors such as **LIDAR**, **ultrasonic sensors**, and **IR sensors**, Arduino-based robots can navigate their environments and make real-time decisions. These robots are also capable of **performing tasks autonomously**, such as sorting objects, assembling products, or even interacting with people.

Example Project:

- **Autonomous Drone**: Arduino can be integrated with drone systems to allow them to fly autonomously. By using an **IMU (Inertial Measurement Unit)** sensor, drones can sense their orientation and adjust their flight path accordingly. This has applications in **delivery systems**, **surveillance**, and **search-and-rescue operations**.

The Future of Arduino in Robotics:

As robotics moves into a new era with the advent of **Industry 4.0**, Arduino will continue to be a driving force in the development of:

- **Collaborative Robots (Cobots)**: These are robots designed to work alongside humans. With Arduino, **cobots** can be programmed to perform tasks like lifting heavy objects, testing products, or assembling parts. They will be more affordable and accessible for small and medium-sized businesses.
- **Robotics in Healthcare**: Arduino is playing a role in the development of **assistive robots**, such as **prosthetics**, **exoskeletons**, and **robotic surgery assistants**. These robots could improve lives, enhance mobility, and aid in medical treatments.
- **Swarm Robotics**: Arduino could also power **swarm robotics**, where a group of simple, inexpensive robots work together to perform complex tasks, like search and rescue or environmental monitoring.

As robotics continues to evolve, Arduino will remain an essential tool for building customizable, scalable, and affordable robots, from personal projects to industrial innovations.

3. Arduino and Smart Cities: Shaping the Urban Future

Smart cities are quickly becoming a global trend, as urban areas seek to become more **sustainable**, **efficient**, and **connected**. The concept of a smart city revolves around integrating **digital technology**, **IoT**, and **big data** to optimize city functions such as transportation, energy use, and public safety. Arduino, as an open-source platform, plays a key role in creating **affordable prototypes** for smart city solutions.

Arduino in Smart Cities Today:

Arduino is already helping make cities smarter by powering numerous IoT devices that can gather data and automate systems in urban environments. For example:

- **Smart Traffic Management**: Arduino systems can control traffic signals based on real-time data from **vehicle sensors** and **CCTV cameras**. These systems can reduce congestion and improve the flow of traffic.
- **Energy Efficiency**: Arduino-powered systems can monitor and control **street lights**, adjusting their brightness based on real-time conditions, such as ambient light levels or traffic flow. This reduces energy consumption.

- **Waste Management**: With **ultrasonic sensors**, Arduino can monitor the fill levels of trash bins and send alerts to waste management teams when they need to be emptied, optimizing waste collection routes and schedules.

Example Project in Smart Cities:

- **Air Quality Monitoring System**: By combining Arduino with **air quality sensors**, it is possible to develop low-cost devices that monitor pollution levels in different parts of a city. These sensors could transmit data to a cloud-based platform, allowing local authorities to take action when pollution exceeds safe levels.

The Future of Arduino in Smart Cities:

As urban areas continue to grow, Arduino's role in smart cities will expand, allowing for the development of:

- **Smart Infrastructure**: Arduino will help design systems that manage **water distribution**, **energy grids**, and **public lighting** more effectively. Arduino-powered systems will be able to optimize energy use, reduce waste, and improve the quality of life for urban residents.
- **Public Safety Systems**: Arduino will enable the creation of low-cost, easy-to-deploy **surveillance cameras**, **sensors** for detecting **gas leaks**, and even **flood detection systems**. These systems can help authorities respond faster to emergencies.
- **Smart Agriculture**: In the context of smart cities, **urban farming** will benefit from Arduino-based

systems that monitor **soil moisture**, **temperature**, and **humidity**. These systems can automate irrigation and improve food production in city environments.

As cities move toward greater **automation**, **efficiency**, and **sustainability**, Arduino will provide the backbone for many smart solutions, enabling cities to become more responsive, livable, and environmentally conscious.

4. Arduino and the Future of Personal Innovation

While AI, robotics, and smart cities dominate the conversation around future technology, Arduino's most profound impact will likely be in empowering individuals to become creators and innovators. Arduino is more than just a development platform—it is a **democratizer of technology**, making powerful tools accessible to anyone with a curious mind and a desire to build.

Personal Innovation and Entrepreneurship

Arduino enables hobbyists, makers, and engineers to develop products that were once out of reach due to cost or technical complexity. The rise of **open-source hardware** means that creators can share and modify designs, leading to an explosion of new ideas and innovations.

Example Project for Personal Innovation:

- **Wearable Tech**: Entrepreneurs can use Arduino to build personal, low-cost wearable devices, such as fitness trackers or heart rate monitors. These projects could form the basis of **startups** that target niche markets in the health, wellness, and fashion industries.

The Future of Personal Innovation with Arduino:

In the future, we can expect Arduino to serve as a launchpad for:

- **Crowdsourced Innovations**: With the rise of platforms like **Kickstarter** and **Indiegogo**, Arduino can empower individuals to prototype and fund their own projects, from **smart glasses** to **personal robots**.
- **Tech for Social Good**: Arduino will continue to be a platform for creating **solutions to global challenges**, such as developing affordable medical devices, disaster relief systems, and educational tools for underserved communities.

By lowering the barrier to entry for hardware development, Arduino will allow more people to innovate, create, and contribute to the technological landscape of tomorrow.

Next-Level Opportunities: Scaling Arduino Projects for Commercialization and Advancing into Mechatronics or Embedded Systems

Arduino is more than just a learning tool; it's a gateway to opportunities in a wide range of industries, from **product development** and **innovation** to **commercialization**. Whether you want to turn your Arduino project into a marketable product or pursue further learning in more advanced fields like **mechatronics** or **embedded systems**, the skills you've developed with Arduino can be the foundation for a rewarding career or business venture. In this section, we'll explore ways to scale your Arduino projects for commercialization and point to advanced fields where you can deepen your expertise.

1. Commercializing Your Arduino Projects: Turning Ideas into Products

As you gain experience and confidence with Arduino, it's natural to think about how you can turn your passion into something more. **Commercialization** involves turning an idea into a product that can be sold to consumers or businesses. While Arduino may start as a prototyping platform, many entrepreneurs use it as a stepping stone to develop fully-fledged commercial products.

How to Scale Your Arduino Project:

1. **Refining Your Prototype**:
    - When your prototype is functional and ready for scaling, it's time to enhance it. This might include upgrading the hardware, improving the design, and making sure the system is robust enough for mass production. Some considerations include:
        - **Power Efficiency**: Ensure the power consumption of your device is optimized, especially if it's meant for battery-powered applications.
        - **Form Factor**: Create a compact, user-friendly design that can be easily marketed.
        - **Durability**: Arduino boards are perfect for prototyping, but for commercial applications, you may need to switch to more rugged or specialized embedded

systems that can handle the demands of manufacturing.

2. **Design for Manufacturability**:
    - Once you've refined your prototype, it's time to consider manufacturing. **PCB design** (Printed Circuit Board) becomes crucial as you move away from Arduino's breadboard-based prototypes.
        - **Design PCBs**: Transition your Arduino setup into a custom PCB design. Tools like **Eagle** or **KiCad** can help you create your own circuit boards, making the product more compact and professional.
        - **Outsource Manufacturing**: You can use online platforms like **PCBWay** or **JLCPCB** to have your custom PCBs manufactured. If the product is mass-market, you may want to engage with a manufacturer to handle larger production runs.
        - **Assembly**: Depending on the complexity of your project, you may need to set up assembly lines or outsource assembly to companies that specialize in small electronics manufacturing.

3. **Product Testing & Certification**:
    - For commercialization, you'll need to conduct **thorough testing** to ensure that your product meets safety and regulatory standards (such as **FCC**, **CE**, or **UL** certifications). For instance, if you're building a **smart home device** or **health monitoring tool**, it's crucial to ensure

that the product meets the required safety and operational standards for consumer use.
4. **Market Research and Marketing**:
    - **Market Research**: Understanding your target audience is key to successful commercialization. Whether your project is focused on **IoT devices**, **wearables**, or **home automation**, conducting research to understand market needs, pricing, and competitor offerings is essential.
    - **Branding and Distribution**: Establish a strong **brand identity** and distribution channels. Platforms like **Kickstarter**, **Indiegogo**, and **Etsy** are popular for launching hardware-based products, while **Amazon** and **eBay** offer a ready-to-go market for selling directly to consumers.
5. **Funding and Partnerships**:
    - Many makers find success in raising funds for commercialization through **crowdfunding platforms** like **Kickstarter**. Alternatively, seeking **investors** or **partnerships** with established businesses in your industry could help scale your product quickly.

Example Project for Commercialization:

- **Smart Plant Care System**: After prototyping a smart irrigation and plant monitoring system with Arduino, you could refine the design by creating a custom PCB for the sensors, integrating it with a mobile app for remote monitoring, and eventually scaling the product for larger consumer markets, like urban gardeners or smart home enthusiasts.

2. Advancing Your Career: Exploring Mechatronics and Embedded Systems

For those interested in deepening their technical skills, **mechatronics** and **embedded systems** are fields where Arduino serves as an excellent starting point. If you've mastered the basics of Arduino, these fields offer the opportunity to expand your knowledge and open up new career opportunities.

Mechatronics: The Intersection of Mechanics, Electronics, and Computing

Mechatronics is a multidisciplinary field that combines mechanical engineering, electronics, computer science, and control engineering to design and create intelligent systems and products. It's particularly relevant for individuals interested in developing advanced **robotics**, **automated systems**, and **smart devices**.

- **Advanced Robotics**: In mechatronics, you'll expand on basic robotics projects by integrating more sophisticated control systems, **sensor fusion**, and **real-time feedback mechanisms**. Arduino can serve as the central brain of a robot, but you might use advanced sensors like **LIDAR**, **GPS**, and **IMUs** for complex navigation tasks.
- **Industrial Automation**: Mechatronics is crucial in industries like **manufacturing**, where automated systems can significantly increase efficiency. Using Arduino's basic principles, you can start building industrial automation systems and gradually explore more complex **PLC** (Programmable Logic Controller) systems used in large factories.

- **Projects in Mechatronics**:
    - **Automated Warehouse Systems**: Imagine designing robots that navigate warehouses, sorting packages and managing inventory. Using Arduino as the prototyping tool, you could integrate advanced control systems and sensors, allowing you to take on real-world challenges.
    - **Medical Robotics**: Another exciting area in mechatronics is medical robotics, where you could develop assistive devices like robotic prosthetics, surgical assistants, or rehabilitation robots.

Embedded Systems: Moving Beyond Arduino to Full-Scale Products

Embedded systems involve designing and programming specialized computers embedded within products or devices. Unlike general-purpose computers, embedded systems are designed to perform specific tasks, such as controlling machinery, monitoring environmental conditions, or enabling **real-time communication**.

- **Microcontroller Platforms**: While Arduino is a fantastic entry point, embedded systems can take you beyond with more powerful **microcontrollers** such as **ARM Cortex**, **Raspberry Pi**, or **ESP32**. These platforms give you more power, memory, and flexibility, enabling you to build more sophisticated projects.
- **Real-Time Operating Systems (RTOS)**: As you progress, learning about **RTOS** can allow you to build more responsive systems, particularly when real-time

performance is crucial—such as in **medical devices** or **automotive systems**.
- **Projects in Embedded Systems**:
    - **Smart Car Systems**: Build advanced systems for autonomous vehicles, integrating sensors and computer vision algorithms.
    - **Consumer Electronics**: Work on smart home devices that require robust embedded systems, such as smart thermostats, security systems, or wearables.

Next Steps for Learning:

- **Advanced Online Courses**: Platforms like **Coursera**, **Udemy**, and **edX** offer courses in mechatronics, embedded systems, and robotics, where you can learn to work with more powerful microcontrollers and advanced hardware platforms.
- **Join Maker Communities and Hackathons**: Getting involved with **open-source hardware communities** and participating in **hackathons** can provide hands-on experience and connect you with professionals in these advanced fields.
- **Read Industry Research Papers**: Stay updated with the latest advancements in mechatronics and embedded systems by reading research papers and attending industry conferences like **IEEE** and **Embedded World**.

3. Invitation to Innovate: Challenging Your Next Big Idea with Arduino

Now that you have the foundational knowledge and insight into potential career paths and commercial opportunities, it's time to push your creativity to new heights. **Arduino's potential is limitless**, and the platform can serve as the starting point for the next big breakthrough.

Challenge Yourself to Innovate:

- **What problem in your daily life can Arduino solve?** Whether it's creating a smart kitchen assistant, a health-monitoring system, or a fully automated garden, Arduino offers a way to turn your ideas into tangible products that can improve everyday life.
- **What's a challenge in your community that needs solving?** Imagine an Arduino-powered device that improves public health, environmental monitoring, or accessibility. Your project could make a real-world impact and even inspire a new business or non-profit initiative.
- **What technologies can you integrate into your Arduino projects?** Think about combining Arduino with **AI**, **IoT**, or **blockchain**. As these technologies continue to evolve, your Arduino-based creations could transform into complex systems that solve new and exciting problems.

Arduino as a Gateway to Limitless Innovation

Arduino provides the perfect blend of accessibility, flexibility, and power, making it the ideal platform for turning ideas into real-world solutions. Whether you're scaling your project for commercialization, diving deeper into mechatronics or embedded systems, or simply brainstorming your next big idea, Arduino empowers you to explore, innovate, and create in ways that were once limited to only the most advanced engineers. Now, it's your turn to take the next step: continue learning, continue building, and let your imagination shape the future of technology.

Conclusion

Embarking on Your Journey with Arduino – A World of Endless Possibilities

Congratulations! By now, you've traveled through the foundational concepts of **Arduino**, from understanding the basics of hardware and software, to building your first projects, exploring sensors and robotics, and troubleshooting like a pro. Along the way, you've not only learned how to bring ideas to life with practical, hands-on skills but also gained the confidence to push the boundaries of what's possible with this incredible technology.

As you close this book, remember that **Arduino** is just the beginning. Whether you're crafting a **simple interactive light**, building **robotic systems**, developing **IoT applications**, or working on **home automation**, you now possess the essential tools and mindset to turn your ideas into real-world creations. But, most importantly, you've learned how to think critically, solve problems, and innovate—skills that are valuable not just for Arduino, but for any technological venture you pursue.

The Power of Practical Experience

By completing the projects and activities in this book, you've demonstrated the power of **practical learning**. Arduino is an approachable platform, but as you move from beginner to intermediate to advanced, you'll notice that the more you experiment, the deeper your understanding becomes. The key to mastery lies not in memorizing instructions, but in **exploring**, **failing**, **iterating**, and **succeeding**. Every line of code you write, every circuit you solder, and every sensor you integrate brings you closer to turning your creative vision into a functioning, innovative product.

Next Steps: Continuing Your Learning Journey

Your journey doesn't stop here. The world of Arduino and electronics is vast, and there is always something new to learn and discover. You've already gained the foundational knowledge to tackle increasingly complex projects, and now it's time to **explore** even deeper areas of technology such as **mechatronics**, **embedded systems**, **artificial intelligence**, and **robotics**. The next step is up to you.

Learning Resources:

To continue learning and developing, there are endless **resources** available to deepen your expertise:

- **Online Communities and Forums**: Platforms like **Arduino's official forum**, **Stack Overflow**, and **Reddit's r/arduino** are great for troubleshooting, collaborating, and gaining insights from other hobbyists and professionals.
- **Advanced Tutorials**: Seek out advanced guides and specialized courses on websites like **Coursera**, **Udemy**, and **edX** to expand your knowledge into more niche areas like **AI integration**, **machine learning**, and **real-time systems**.
- **Hackathons and Maker Fairs**: Engaging with **community-driven events** like hackathons or maker fairs is an excellent way to test your skills, collaborate with others, and showcase your projects to potential collaborators or investors.

From Prototyping to Product Creation

You've already seen how Arduino can serve as a **tool for innovation**, and as you build more advanced projects, you'll find that the skills you've developed can be leveraged for **commercialization**. Whether you dream of turning your **Arduino-based inventions** into products or simply want to create for the joy of it, this book has provided the framework for turning abstract ideas into functional, tangible results. With more practice, your ability to design, prototype, and refine complex systems will grow, opening doors to future business ventures or career opportunities in the **tech industry**.

Final Thoughts: Embrace Creativity and Innovation

Ultimately, Arduino is not just about circuits and code; it's about **creativity**, **innovation**, and the drive to transform ideas into reality. As you continue your journey, remember that there are no limits to what you can achieve. Whether you're building a **smart city solution**, an **advanced robotics system**, or simply creating an innovative personal project, Arduino provides the perfect platform for turning your imagination into something concrete.

The world of **Arduino** is vibrant and evolving, and you, as a creator, are part of that evolution. As you experiment, innovate, and collaborate, keep pushing the boundaries of your knowledge and capabilities. Your next project could be the one that changes the way people interact with the world, the one that leads to breakthroughs in technology, or the one that sparks your next big idea.

The future is yours to build.

So, go ahead—build something amazing with Arduino. The possibilities are endless.

References

Below are some essential resources to further deepen your understanding of Arduino programming, electronics, robotics, and coding. These references will guide you in continuing your journey and exploring new technologies as you develop your skills:

1. **Arduino Official Documentation**
   Arduino. (n.d.). *Arduino Reference*. Retrieved from https://www.arduino.cc/reference/en/
   The official documentation provides comprehensive information on the Arduino programming language, functions, and board specifications.
2. **Getting Started with Arduino**
   Monk, S. (2016). *Programming Arduino: Getting Started with Sketches* (2nd ed.). McGraw-Hill Education.
   A beginner-friendly guide for understanding the basics of Arduino programming and hardware, perfect for those new to electronics and coding.
3. **Arduino Cookbook**
   Purdie, M., & DeMauro, D. (2014). *Arduino Cookbook: Recipes to Begin, Expand, and Enhance Your Projects* (2nd ed.). O'Reilly Media.
   A comprehensive collection of Arduino recipes that provide solutions for common hardware and software challenges in various Arduino projects.
4. **Exploring Arduino: Tools and Techniques for Engineering Wizardry**
   Margolis, M. (2011). *Exploring Arduino: Tools and Techniques for Engineering Wizardry*. Wiley.
   This book dives deeper into Arduino's capabilities,

offering more advanced projects and applications for those looking to take their skills further.

5. **The Arduino Programmer's Guide**
   Eberle, M. (2013). *Arduino: A Technical Reference.* O'Reilly Media.
   A reference for more advanced programmers, this guide covers Arduino's inner workings and provides insights into optimizing projects for efficiency.

6. **Make: Electronics**
   Charles Platt. (2009). *Make: Electronics: Learning Through Discovery* (2nd ed.). Maker Media.
   A great introduction to electronics for beginners, this book uses hands-on experiments to help readers learn the basics of circuits, components, and how they interact with microcontrollers like Arduino.

7. **The Art of Electronics**
   Horowitz, P., & Hill, W. (2015). *The Art of Electronics* (3rd ed.). Cambridge University Press.
   A deep dive into electronics theory, offering a more comprehensive understanding of components and circuits for Arduino projects.

8. **Arduino for Robots and Peripherals**
   O'Reilly, R. (2011). *Arduino Robotics.* McGraw-Hill Professional.
   This book focuses on how to integrate Arduino into robot-building projects, from simple prototypes to more complex robots, ideal for those interested in the intersection of robotics and Arduino.

9. **Hackaday**
   Hackaday. (n.d.). *Arduino Projects and Tutorials.* Retrieved from
   https://hackaday.com/category/arduino/
   Hackaday is a treasure trove of DIY projects and

tutorials, many centered on Arduino. It's a great community for inspiration and innovation.

10. **Instructables**
    Instructables. (n.d.). *Arduino Projects*. Retrieved from https://www.instructables.com/howto/arduino/
    A community-driven platform where users share step-by-step instructions for Arduino-based projects, ranging from beginner to expert-level.

11. **The Official Arduino Forum**
    Arduino Forum. (n.d.). *Arduino Community*. Retrieved from https://forum.arduino.cc/
    An official online forum where hobbyists, makers, and professionals alike discuss projects, troubleshooting, and advancements in the Arduino world.

12. **Arduino Project Hub**
    Arduino. (n.d.). *Arduino Project Hub*. Retrieved from https://create.arduino.cc/projecthub
    A platform where users can publish their own Arduino projects and explore other people's creations, providing both inspiration and practical tips.

13. **Designing Embedded Systems with Arduino**
    Hibbard, J. (2012). *Designing Embedded Systems with Arduino*. McGraw-Hill Education.
    A resource for developing more sophisticated embedded systems and a great follow-up for readers ready to move beyond basic projects.

14. **Raspberry Pi and Arduino Integration**
    Monk, S. (2015). *Raspberry Pi and Arduino: Building a Smart Home* (1st ed.). McGraw-Hill Education.
    Learn how to integrate both Raspberry Pi and Arduino in creating smart home systems and Internet of Things (IoT) devices.

15. **GitHub - Arduino Repositories**
    GitHub. (n.d.). *Arduino Repositories*. Retrieved from

https://github.com/arduino

The official Arduino GitHub repository, where you can explore thousands of open-source Arduino projects, libraries, and code examples shared by the community.

16. **Hackaday.io**

    Hackaday. (n.d.). *Hackaday.io Projects*. Retrieved from https://hackaday.io/

    Another excellent resource for open-source hardware projects, Hackaday.io offers insights from other makers working with Arduino and similar platforms.

By tapping into these references, you'll not only continue to expand your knowledge but also stay at the forefront of Arduino innovation. The possibilities are vast, and with the tools, knowledge, and resources provided in this book, you're well-equipped to take on whatever Arduino-based projects come your way.

Happy building, and may your creativity continue to thrive!

About the Author

Maxwell Harper is a passionate engineer, educator, and innovator with a deep love for technology, electronics, and teaching others how to bring their ideas to life. With a background in electrical engineering and a decade of hands-on experience in the fields of robotics, IoT (Internet of Things), and embedded systems, Maxwell has worked on everything from small-scale DIY projects to large, industry-driven technological solutions.

Maxwell's journey with Arduino began as a way to simplify the often-intimidating world of electronics, and over the years, it has evolved into a mission to empower individuals of all ages and backgrounds to unlock their creative potential through coding and technology. Maxwell has taught Arduino workshops globally, helping students—from beginners to advanced makers—navigate the world of electronics, design, and prototyping.

In addition to his work as an educator, Maxwell has been involved in several groundbreaking open-source projects, contributing code and collaborating with a global community of developers. His passion for accessible learning shines through in his books and tutorials, which focus on simplifying complex concepts and encouraging hands-on experimentation. Maxwell is an advocate for lifelong learning and believes that anyone, regardless of age or experience, can become a maker and innovate using the tools of tomorrow.

When he's not designing new projects or writing tutorials, Maxwell can be found speaking at tech conferences, running hackathons, and mentoring aspiring engineers and creators.

Disclaimer:

The content of this book is intended for informational and educational purposes only. While every effort has been made to ensure the accuracy of the information provided, the author and publisher make no representations or warranties regarding the completeness, accuracy, or reliability of the material. The information and projects in this book are intended for use by individuals who are interested in learning about Arduino programming, robotics, and related technologies. Readers should exercise caution and follow appropriate safety procedures when working with electronics, components, and electrical equipment.

The author and publisher disclaim any liability for any direct, indirect, incidental, or consequential damages arising from the use or inability to use the information, code, or projects provided in this book. By using the content in this book, you agree to take full responsibility for your own actions and any risks associated with the implementation of the projects.

Additionally, the author is not responsible for any damage to personal property, hardware, or software that may occur as a result of following the instructions in this book.

This book is not affiliated with or endorsed by the official Arduino project or any of its partners or affiliates. All trademarks, registered trademarks, and product names mentioned in this book are the property of their respective owners.

Always consult relevant manuals, official documentation, and experts when handling electronics and programming projects.

Copyright © 2024 Maxwell Harper. All rights reserved.

No part of this book may be reproduced, distributed, or transmitted in any form or by any means, including photocopying, recording, or other electronic or mechanical methods, without the prior written permission of the author, except in the case of brief quotations embodied in critical reviews or articles. For permission requests, please contact the author.

This book is a work of nonfiction. Any references to real people, events, or organizations are used in a factual manner for the purpose of illustration and education. All trademarks, service marks, and product names mentioned in this book are the property of their respective owners.

The information contained in this book is for educational purposes only, and the author makes no claims regarding the effectiveness or safety of the projects or activities described. Readers are encouraged to follow safety guidelines and consult additional resources when working with electronics, programming, or related materials.

www.ingramcontent.com/pod-product-compliance
Lightning Source LLC
Chambersburg PA
CBHW052247220526
45471CB00001B/222